James Alexander McLellan, John Dewey

The Psychology of Number and Its Applications to Methods of Teaching Arithmetic

James Alexander McLellan, John Dewey

The Psychology of Number and Its Applications to Methods of Teaching Arithmetic

ISBN/EAN: 9783337168049

Printed in Europe, USA, Canada, Australia, Japan

Cover: Foto ©Andreas Hilbeck / pixelio.de

More available books at **www.hansebooks.com**

International Education Series

EDITED BY

WILLIAM T. HARRIS, A. M., LL. D.

VOLUME XXXVIII.

THE INTERNATIONAL EDUCATION SERIES.

12mo, cloth, uniform binding.

THE INTERNATIONAL EDUCATION SERIES was projected for the purpose of bringing together in orderly arrangement the best writings, new and old, upon educational subjects, and presenting a complete course of reading and training for teachers generally. It is edited by W. T. HARRIS, LL. D., United States Commissioner of Education, who has contributed for the different volumes in the way of introductions, analysis, and commentary. The volumes are tastefully and substantially bound in uniform style.

VOLUMES NOW READY.

Vol. I.—THE PHILOSOPHY OF EDUCATION. By JOHANN K. F. ROSENKRANZ, Doctor of Theology and Professor of Philosophy, University of Königsberg. Translated by ANNA C. BRACKETT. Second edition, revised, with Commentary and complete Analysis. $1.50.

Vol. II.—A HISTORY OF EDUCATION. By F. V. N. PAINTER, A. M., Professor of Modern Languages and Literature, Roanoke College, Va. $1.50.

Vol. III.—THE RISE AND EARLY CONSTITUTION OF UNIVERSITIES. WITH A SURVEY OF MEDIÆVAL EDUCATION. By S. S. LAURIE, LL. D., Professor of the Institutes and History of Education, University of Edinburgh. $1.50.

Vol. IV.—THE VENTILATION AND WARMING OF SCHOOL BUILDINGS. By GILBERT B. MORRISON, Teacher of Physics and Chemistry, Kansas City High School. $1.00.

Vol. V.—THE EDUCATION OF MAN. By FRIEDRICH FROEBEL. Translated and annotated by W. N. HAILMANN, A. M., Superintendent of Public Schools, La Porte, Ind. $1.50.

Vol. VI.—ELEMENTARY PSYCHOLOGY AND EDUCATION. By JOSEPH BALDWIN, A. M., LL. D., author of "The Art of School Management." $1.50.

Vol. VII.—THE SENSES AND THE WILL. (Part I of "THE MIND OF THE CHILD.") By W. PREYER, Professor of Physiology in Jena. Translated by H. W. BROWN, Teacher in the State Normal School at Worcester, Mass. $1.50.

Vol. VIII.—MEMORY: WHAT IT IS AND HOW TO IMPROVE IT. By DAVID KAY, F. R. G S., author of "Education and Educators," etc. $1.50.

Vol. IX.—THE DEVELOPMENT OF THE INTELLECT. (Part II of "THE MIND OF THE CHILD.") By W. PREYER, Professor of Physiology in Jena. Translated by H. W. BROWN. $1.50.

Vol. X.—HOW TO STUDY GEOGRAPHY. A Practical Exposition of Methods and Devices in Teaching Geography which apply the Principles and Plans of Ritter and Guyot. By FRANCIS W. PARKER, Principal of the Cook County (Illinois) Normal School. $1.50.

Vol. XI.—EDUCATION IN THE UNITED STATES: ITS HISTORY FROM THE EARLIEST SETTLEMENTS. By RICHARD G. BOONE, A. M., Professor of Pedagogy, Indiana University. $1.50.

Vol. XII.—EUROPEAN SCHOOLS: OR, WHAT I SAW IN THE SCHOOLS OF GERMANY, FRANCE, AUSTRIA, AND SWITZERLAND. By L. R. KLEMM, Ph. D., Principal of the Cincinnati Technical School. Fully Illustrated. $2.00.

THE INTERNATIONAL EDUCATION SERIES.—(Continued.)

Vol. XIII.—PRACTICAL HINTS FOR THE TEACHERS OF PUBLIC SCHOOLS. By GEORGE HOWLAND, Superintendent of the Chicago Public Schools. $1.00.

Vol. XIV.—PESTALOZZI: HIS LIFE AND WORK. By ROGER DE GUIMPS. Authorized Translation from the second French edition, by J. RUSSELL, B. A. With an Introduction by Rev. R. H. QUICK, M. A. $1.50.

Vol. XV.—SCHOOL SUPERVISION. By J. L. PICKARD, LL. D. $1.00.

Vol. XVI.—HIGHER EDUCATION OF WOMEN IN EUROPE. By HELENE LANGE, Berlin. Translated and accompanied by comparative statistics by L. R. KLEMM. $1.00.

Vol. XVII.—ESSAYS ON EDUCATIONAL REFORMERS. By ROBERT HERBERT QUICK, M. A., Trinity College, Cambridge. *Only authorized edition of the work as rewritten in 1890.* $1.50.

Vol. XVIII.—A TEXT-BOOK IN PSYCHOLOGY. By JOHANN FRIEDRICH HERBART. Translated by MARGARET K. SMITH. $1.00.

Vol. XIX.—PSYCHOLOGY APPLIED TO THE ART OF TEACHING. By JOSEPH BALDWIN, A. M., LL. D. $1.50.

Vol. XX.—ROUSSEAU'S ÉMILE; OR, TREATISE ON EDUCATION. Translated and annotated by W. H. PAYNE, Ph. D., LL. D., Chancellor of the University of Nashville. $1.50.

Vol. XXI.—THE MORAL INSTRUCTION OF CHILDREN. By FELIX ADLER. $1.50.

Vol. XXII.—ENGLISH EDUCATION IN THE ELEMENTARY AND SECONDARY SCHOOLS. By ISAAC SHARPLESS, LL. D., President of Haverford College. $1.00.

Vol. XXIII.—EDUCATION FROM A NATIONAL STANDPOINT. By ALFRED FOUILLÉE. $1.50.

Vol. XXIV.—MENTAL DEVELOPMENT IN THE CHILD. By W. PREYER, Professor of Physiology in Jena. Translated by H. W. BROWN. $1.00.

Vol. XXV.—HOW TO STUDY AND TEACH HISTORY. By B. A. HINSDALE, Ph. D., LL. D., University of Michigan. $1.50.

Vol. XXVI.—SYMBOLIC EDUCATION: A COMMENTARY ON FROEBEL'S "MOTHER PLAY." By SUSAN E. BLOW. $1.50.

Vol. XXVII.—SYSTEMATIC SCIENCE TEACHING. By EDWARD GARDNIER HOWE. $1.50.

Vol. XXVIII.—THE EDUCATION OF THE GREEK PEOPLE. By THOMAS DAVIDSON. $1.50.

Vol. XXIX.—THE EVOLUTION OF THE MASSACHUSETTS PUBLIC-SCHOOL SYSTEM. By G. H. MARTIN, A. M. $1.50.

Vol. XXX.—PEDAGOGICS OF THE KINDERGARTEN. By FRIEDRICH FROEBEL. 12mo. $1.50.

Vol. XXXI.—THE MOTTOES AND COMMENTARIES OF FRIEDRICH FROEBEL'S MOTHER PLAY. By HENRIETTA R. ELIOT and SUSAN E. BLOW.

Vol. XXXII.—THE SONGS AND MUSIC OF FRIEDRICH FROEBEL'S MOTHER PLAY (MUTTER UND KOSE LIEDER). By SUSAN E. BLOW.

New York: D. APPLETON & CO., Publishers, 72 Fifth Avenue.

THE
PSYCHOLOGY OF NUMBER

AND ITS APPLICATIONS TO
METHODS OF TEACHING ARITHMETIC

BY

JAMES A. McLELLAN, A. M., LL. D.
PRINCIPAL OF THE ONTARIO SCHOOL OF PEDAGOGY, TORONTO

AND

JOHN DEWEY, Ph. D.
HEAD PROFESSOR OF PHILOSOPHY IN THE UNIVERSITY OF CHICAGO

"The art of measuring brings the world into subjection to man; the art of writing prevents his knowledge from perishing along with himself; together they make man—what Nature has not made him—all-powerful and eternal."—MOMMSEN.

NEW YORK
D. APPLETON AND COMPANY
1895

COPYRIGHT, 1895,
BY D. APPLETON AND COMPANY.

ELECTROTYPED AND PRINTED
AT THE APPLETON PRESS, U. S. A.

EDITOR'S PREFACE.

In presenting this book on the Psychology of Number it is believed that a special want is supplied. There is no subject taught in the elementary schools that taxes the teacher's resources as to methods and devices to a greater extent than arithmetic. There is no subject taught that is more dangerous to the pupil in the way of deadening his mind and arresting its development, if bad methods are used. The mechanical side of training must be joined to the intellectual in such a form as to prevent the fixing of the mind in thoughtless habits. While the mere processes become mechanical, the mind should by ever-deepening insight continually increase its power to grasp details in more extensive combinations.

Methods must be chosen and justified, if they can be justified at all, on psychological grounds. The concept of number will at first be grasped by the pupil imperfectly. He will see some phases of it and neglect others. Later on he will arrive at operations which demand a view of all that number implies. Each and every number is an implied ratio, but it does not express the ratio as simple number. The German language is fortunate in having terms that express the two aspects of numerical quantity. *Anzahl* expresses the multiplicity and *Einheit* the unity. Any number, say six, for example, has these two aspects: it is a manifold of units; the constituent unit whatever it is, is repeated six times. It is a unity of these, and as such may be a constituent unit of a larger number, five times six, for instance, wherein the five represents the multiplicity (*Anzahl*) and the six the constituent unity (*Einheit*).*

Number is one of the developments of quantity. Its multiplicity and unity correspond to the two more general aspects of quantity in general, namely, to discreteness and continuity.

There is such a thing as qualitative unity, or individuality. Quantitative unity, unlike individuality, is always divisible into constituent units. All quantity is a unity of units. It is composed of constituent units, and it is itself a constituent unit of a real or possible larger unity. Every pound contains within it ounces; every pound is a constituent unit of some hundredweight or ton.

The simple number implies both phases, the multiplicity and the unity, but does not express them adequately. The child's thought likewise possesses the same inadequacy; it implies more than it explicitly states or holds in consciousness.

This twofoldness of number becomes explicit in multiplication and division, wherein one number is the unit and the other expresses the multiplicity—the *times* the unit is taken. Fractions form a more adequate expression of this ratio, and require a higher consciousness of the nature of quantity than simple numbers do. Hence the difficulty of teaching this subject in the ele-

---

* Hegel, Logik, Bd. I, 1st Th., S. 225.

mentary school. The thought of $\frac{7}{8}$ demands the thought of both numbers, 7 and 8, and the thought of their modification each through the other.

The methods in vogue in elementary schools are chiefly based on the idea that it is necessary to eliminate the ratio idea by changing one of the terms of the fraction to a qualitative unit and by this to change the thought to that of a simple number. Thus halves and quarters and cents and dimes are thought as individual things, and the fractional idea suppressed.

In the differential calculus ratio is most adequately expressed as the fundamental and true form of all quantity, number included. The differential of x and the differential of y are ratios.

The authors of this book have presented in an admirable manner this psychological view of number, and shown its application to the correct methods of teaching the several arithmetical processes. The shortcomings of the "fixed-unit" theory are traced out in all their consequences. The defects of a view which makes unity a qualitative instead of a quantitative idea are sure to appear in the methods of solution adopted.

Pupils studying music by the highest method learn thoroughly those combinations which involve double counterpoint. As soon as the hands are trained to readily execute such exercises the pupil can take up a sonata of Beethoven or a fugue of Bach, and soon become familiar with it. On the plan of the old lessons in counterpoint, the pupil found himself helpless before such a composition. His phrases furnished no key to the compositions of Bach or Beethoven, because the latter are constructed on a different counterpoint.

So the methods of teaching arithmetic by a "fixed-unit" system do not lead towards the higher mathematics, but away from it. They furnish little, if any, training in thinking the ratio involved in the very idea of number.

The psychology of number requires that the methods be chosen with reference to their power to train the mind of the pupil into this consciousness of the ratio idea. The steps should be short and the ascent gradual; but it should be continuous, so that the pupil constantly gains in his ability to hold in consciousness the unity of the two aspects of quantity, the unity of the discrete and the continuous, the unity of the multiplex and the simple unit.

Measurement is a process that makes these elements clear. The constituent unit becomes the including unit, and *vice versa*, through being measured and being made the measure of others. This, too, is involved in using the decimal system of numeration, and in understanding the different orders of units, each of which both includes constituent units and is included as a constituent of a higher unit.

The hint is obtained from this that the first lessons in arithmetic should be based on the practice of measuring in its varied applications.

Again, since ratio is the fundamental idea, one sees how fallacious are those theories which seek to lay a basis for mathematics by at first producing a clear and vivid idea of unity—as though the idea of quantity were to be built up on this idea. It is shown that such abstract unit is not yet quantity nor an element of quantity, but simply the idea of individuality, which is still a qualitative idea, and does not become quantitative until it is conceived as composite and made up of constituent units homogeneous with itself.

The true psychological theory of number is the panacea for that exaggeration of the importance of arithmetic which prevails in our elementary schools. As if it were not enough that the science of number is indispensable for the conquest of Nature in time and space, these qualitative-unit teachers make the mistake of supposing that arithmetic deals with spiritual being as much as with matter; they confound quality with quantity, and consequently mathematics with metaphysics. Mental arithmetic becomes in their psychology "the discipline for the pure reason," although as a matter of fact the three figures of the regular syllogism are neither of them employed in mathematical reasoning.

W. T. HARRIS.

WASHINGTON, D. C., *June 25, 1895.*

PREFACE.

It is perhaps natural that a growing impatience with the meagre results of the time given to arithmetic in the traditional course of the schools should result in attacks upon that study. While not all educational experts would agree that it is the " most useless of all subjects " taught, there is an increasing tendency to think of it and speak of it as a necessary evil, and therefore to be kept within the smallest possible bounds. However natural this reaction, it is none the less unwise when turned against arithmetic itself, and not against stupid and stupefying ways of teaching it. So conceived, the movement stands only for an aimless swing of the scholastic pendulum, sure to be followed by an equally unreasonable swing to the other extreme. If methods which cut across the natural grain of the mental structure and resist the straightforward workings of the mental machinery, waste time, create apathy and disgust, dull the power of quick perception, and cultivate habits of inaccurate and disconnected attention, what occasion for surprise? Because wrong methods breed bad results, it hardly follows that education can be made symmetrical by omitting a subject which stands *par excellence* for clear and clean cut methods of thought, which forms the introduction to all effective interpretation of Nature, and is a powerful instrument in the regulation of social intercourse.

It is customary now to divide studies into "form" studies and "content" studies, and to depreciate arithmetic on the ground that it is merely formal. But how are we to separate form and content, and regard one as good in itself and the other as, at best, a necessary evil? If we may paraphrase a celebrated saying of Kant's, while form without content is barren, content without form is mushy. An education which neglects the formal relationships constituting the framework of the subject-matter taught is inert and supine. The pedagogical problem is not solved by railing at "form," but in discovering what kind of form we are dealing with, how it is related to its own content, and in working out the educational methods which answer to this relationship. Because, in the case of number, "form" represents the measured adjustment of means to an end, the rhythmical balancing of parts in a whole, the mastery of form represents directness, accuracy, and economy of perception, the power to discriminate the relevant from the irrelevant, and ability to mass and converge relevant material upon a destined end—represents, in short, precisely what we understand by good sense, by good judgment, the power to put two and two together. When taught as this sort of form, arithmetic affords in its own place an unrivalled means of mental discipline. It is, perhaps, more than a coincidence that the particular school of educational thought which is most active in urging the merely "formal" quality of arithmetic is also the one which stands most systematically for what is condemned in the following pages as the "fixed unit" method of teaching.

As for the counterpart objection that number work is lacking in ethical substance and stimulus, much may be learned from a study of Greek civilization, from the recognition of the part which Greek theory and practice assigned to the ideas of rhythm, of balance, of measure, in moral and æsthetic culture. That the Greeks also kept their arithmetical training in closest connection with the study of spatial forms, with measurement, may again be more than a coincidence. Even upon its merely formal side, a study which requires exactitude, continuity, patience, which automatically rejects all falsification of data, all slovenly manipulation, which sets up a controlling standard of balance at every point, can hardly be condemned as lacking in the ethical element. But this idea of balance, of compensation, is more than formal. Number represents, as is shown in the following pages, *valuation;* number is the tool whereby modern society in its vast and intricate processes of exchange introduces system, balance and economy into those relationships upon which our daily life depends. Properly conceived and presented, neither geography nor history is a more effective mode of bringing home to the pupil the realities of the social environment in which he lives than is arithmetic. Society has its form also, and it is found in the processes of fixing standards of value and methods of valuation, the processes of weighing and counting, whether distance, size, or quality; of measuring and fixing bounds, whether in space or time, and of balancing the various resulting values against one another. Arithmetic can not be properly taught without being an introduction to this form.

Thanks are due to Professor Scott, of the Toronto Normal School, for some assistance with the proofs; to Principal Harston, of the Collegiate Institute, for some excellent solutions; and to Mr. De Lury, of University College, Lecturer on Methods in Mathematics in Ontario School of Pedagogy, for valued practical assistance.

August 13, 1895.

CONTENTS.

	PAGE
Editor's Preface	v
Author's Preface	xi

CHAPTER
I.—What Psychology can do for the Teacher . .	1
II.—The Psychical Nature of Number	23
III.—The Origin of Number: Dependence of Number on Measurement, and of Measurement on Adjustment of Activity	35
IV.—The Origin of Number: Summary and Applications	52
V.—The Definition, Aspects, and Factors of Numerical Ideas	68
VI.—The Development of Number: or, the Arithmetical Operations	93
VII.—Numerical Operations as External and as Intrinsic to Number	119
VIII.—On Primary Number Teaching	144
IX.—On Primary Number Teaching	166
X.—Notation, Addition, Subtraction	190
XI.—Multiplication and Division	207
XII.—Measures and Multiples	227
XIII.—Fractions	241
XIV.—Decimals	261
XV.—Percentage and its Applications . . .	279
XVI.—Evolution	297

2

THE PSYCHOLOGY OF NUMBER.

CHAPTER I.

WHAT PSYCHOLOGY CAN DO FOR THE TEACHER.

The value of any fact or theory as bearing on human activity is, in the long run, determined by practical application—that is, by using it for accomplishing some definite purpose. If it works well—if it removes friction, frees activity, economizes effort, makes for richer results—it is valuable as contributing to a perfect adjustment of means to end. If it makes no such contribution it is practically useless, no matter what claims may be theoretically urged in its behalf. To this the question of the relation between psychology and education presents no exception. The value of a knowledge of psychology in general, or of the psychology of a particular subject, will be best made known by its fruits. No amount of argument can settle the question once for all and in advance of any experimental work. But, since education is a rational process, that is a process in harmony with the laws of psychical development, it is plain that the educator need not and should not depend upon vague inductions from a practice not grounded upon principles. Psychology can not dispense with experience, nor can experience, if it is to be rational, dispense with psychology. It is possible to make actual practice less a matter of mere experiment and more a matter of reason; to make it contribute directly and economically to a rich and ripe, because rational, experience. And this the educational psychologist attempts to do by indicating in what directions help is likely to be found; by indicating what kind of psychology is likely to help and what is not likely; and, finally, by indicating what valid reasons there are for anticipating any help at all.

I. As to the last point suggested, that psychology *ought* to help the educator, there can be no disagreement. In the *first place* the study of psychology has a high *disciplinary* value for the teacher. It develops the power of connected thinking and trains to logical habits of mind. These qualities, essential though they are in thorough teaching, there is a tendency to undervalue in educational methods of the present time when so much is made of the accumulation of facts and so little of their organization. In our eager advocacy of "facts and things" we are apparently forgetting that these are comparatively worthless, either as stored knowledge or for developing power, till they have been subjected to the discriminating and formative energy of the intelligence. Unrelated facts are not knowledge any more than the words of a dictionary are connected thoughts. And so the work of getting "things" may be carried to such an extent as to burden the mind and check the growth of its higher powers. There may be a surfeit of things with the usual consequence of an impaired mental digestion. It is pretty generally conceded that the number of facts memorized is by no means a measure of the amount of power developed; indeed, unless reflection has been exercised step by step with observation, the mass of power gained may turn out to be inversely proportional to the multitude of facts. This does not mean that there is any opposition between reflection and true observation. There can not be observation in the best sense of the word without reflection, nor can reflection fail to be an effective preparation for observation.

It will be readily admitted that this tendency to exalt facts unduly may be checked by the study of psychology. Here, in a comparatively abstract science, there *must* be reflection—abstraction and generalization. In nature study we gather the facts, and we *may* reflect upon the facts: in mind study we must reflect in order to get the facts. To observe the subtle and complex facts of mind, to discriminate the elements of a consciousness never the same for two successive moments, to give unity of meaning to these abstract mental phenomena, demands such concentration of attention as must secure the growth of mental power—power to master, and not be mastered by, the facts and ideas of whatever kind which may be crowding in upon the mind; to resolve a complex subject into its component parts, seizing upon the most important and holding them clearly defined and related in consciousness; to take, in a word, any "chaos" of experience and reduce it to harmony and system. This analytic and relating power, which is an essential mark of the clear thinker, is the prime qualification of the clear teacher.

But, in the *second place*, the study of psychology is

of still more value to the teacher in its bearing upon his *practical* or strictly professional training.

Every one grants that the primary aim of education is the training of the powers of intelligence and will—that the object to be attained is a certain quality of character. To say that the purpose of education is "an increase of the powers of the mind rather than an enlargement of its possessions"; that education is a science, the science of the formation of character; that character means a measure of mental power, mastery of truths and laws, love of beauty in nature and in art, strong human sympathy, and unswerving moral rectitude; that the teacher is a trainer of mind, a former of character; that he is an artist above nature, yet in harmony with nature, who applies the science of education to help another to the full realization of his personality in a character of strength, beauty, freedom—to say this is simply to proclaim that the problem of education is essentially an ethical and psychological problem. This problem can be solved only as we know the true nature and destination of man as a rational being, and the rational methods by which the perfection of his nature may be realized. Every aim proposed by the educator which is not in harmony with the intrinsic aim of human nature itself, every method or device employed by the teacher that is not in perfect accord with the mind's own workings, not only wastes time and energy, but results in positive and permanent harm; running counter to the true activities of the mind, it certainly distorts and may possibly destroy them. To the educator, therefore, the only solid ground of assurance that he is not setting up impossible or artificial aims, that he is not using ineffective and perverting methods, is a clear and definite knowledge of the normal end and the normal forms of mental action. To know these things is to be a true psychologist and a true moralist, and to have the essential qualifications of the true educationist. Briefly, only psychology and ethics can take education out of its purely empirical and rule-of-thumb stage. Just as a knowledge of mathematics and mechanics has wrought marvelous improvements in all the arts of construction; just as a knowledge of steam and electricity has made a revolution in modes of communication, travel, and transportation of commodities; just as a knowledge of anatomy, physiology, pathology has transformed medicine from empiricism to applied science, so a knowledge of the structure and functions of the human being can alone elevate the school from the position of a mere workshop, a more or less cumbrous, uncertain, and even baneful institution, to that of a vital, certain, and effective instrument in the greatest of all constructions—the building of a free and powerful character.

Without the assured methods and results of science there are just three resources available in the work of education.

1. The first is *native tact and skill*, the intuitive power that comes mainly from sympathy. For this personal power there is absolutely no substitute. "Any one can keep school," perhaps, but not every one can teach school any more than every one can become a capable painter, or an able engineer, or a skilled artist in any direction. To ignore native aptitude, and to depend wholly, or even chiefly, upon the knowledge and

use of "methods," is an error fatal to the best interests of education; and there can be no question that many schools are suffering frightfully from ignoring or undervaluing this paramount qualification of the true teacher. But in urging the need of psychology in the preparation of the teacher there is no question of ignoring personal power or of finding a substitute for personal magnetism. It is only a question of providing the best opportunities for the exercise of native capacity—for the fullest development and most fruitful application of endowments of heart and brain. Training and native outfit, culture and nature, are never opposed to each other. It is always a question, not of suppressing or superseding, but of cultivating native instinct, of training natural equipment to its ripest development and its richest use. A Pheidias does not despise learning the principles necessary to the mastery of his art, nor a Beethoven disregard the knowledge requisite for the complete technical skill through which he gives expression to his genius. In a sense it is true that the great artist is born, not made; but it is equally true that a scientific insight into the technics of his art *helps* to make him. And so it is with the artist teacher. The greater and more scientific his knowledge of human nature, the more ready and skilful will be his application of principles to varying circumstances, and the larger and more perfect will be the product of his artistic skill.

But the genius in education is as rare as the genius in other realms of human activity. Education is, and forever will be, in the hands of ordinary men and women; and if psychology—as the basis of scientific insight into human nature—is of high value to the few who possess genius, it is indispensable to the many who have not genius. Fortunately for the race, most persons, though not "born" teachers, are endowed with some "genial impulse," some native instinct and skill for education; for the cardinal requisite in this endowment is, after all, sympathy with human life and its aspirations. We are all born to be educators, to be parents, as we are not born to be engineers, or sculptors, or musicians, or painters. Native capacity for education is therefore much more common than native capacity for any other calling. Were it not so, human society could not hold together at all. But in most people this native sympathy is either dormant or blind and irregular in its action; it needs to be awakened, to be cultivated, and above all to be intelligently directed. The instinct to walk, to speak, and the like are imperious instincts, and yet they are not wholly left to "nature"; we do not assume that they will take care of themselves; we stimulate and guide, we supply them with proper conditions and material for their development. So it must be with this instinct, so common yet at present so comparatively ineffective, which lies at the heart of all educational efforts, the instinct to help others in their struggle for self-mastery and self-expression. The very fact that this instinct is so strong, and all but universal, and that the happiness of the individual and of the race so largely depends upon its development and intelligent guidance, gives greater force to the demand that its growth may be fostered by favourable conditions; and that it may be made certain and reasonable in its action, instead of being left blind and faltering, as it surely will be without rational cultivation.

To this it may be added that native endowment can work itself out in the best possible results only when it works under right conditions. Even if scientific insight were not a necessity for the true educator himself, it would still remain a necessity for others in order that they might not obstruct and possibly drive from the profession the teacher possessed of the inborn divine light, and restrict or paralyze the efforts of the teacher less richly endowed. It is the mediocre and the bungler who can most readily accommodate himself to the conditions imposed by ignorance and routine; it is the higher type of mind and heart which suffers most from its encounter with incapacity and ignorance. One of the greatest hindrances to true educational progress is the reluctance of the best class of minds to engage in educational work precisely because the general standard of ethical and psychological knowledge is so low that too often high ideals are belittled and efforts to realize them even vigorously opposed. The educational genius, the earnest teacher of any class, has little to expect from an indifference, or a stolidity, which is proof alike against the facts of experience and the demonstrations of science.

2. The Second Resource is *experience*. This, again, is necessary. Psychology is not a short and easy path that renders personal experience superfluous. The real question is: What kind of experience shall it be? It is in a way perfectly true that only by teaching can one become a teacher. But not any and every sort of thing which passes for teaching or for "experience" will

make a teacher any more than simply sawing a bow across violin strings will make a violinist. It is a certain quality of practice, not mere practice, which produces the expert and the artist. Unless the practice is based upon rational principles, upon insight into facts and their meaning, "experience" simply fixes incorrect acts into wrong habits. Nonscientific practice, even if it finally reaches sane and reasonable results—which is very unlikely—does so by unnecessarily long and circuitous routes; time and energy are wasted that might easily be saved by wise insight and direction at the outset.

The worst thing about empiricism in every department of human activity is that it leads to a blind observance of rule and routine. The mark of the empiric is that he is helpless in the face of new circumstances; the mark of the scientific worker is that he has power in grappling with the new and the untried; he is master of principles which he can effectively apply under novel conditions. The one is a slave of the past, the other is a director of the future. This attachment to routine, this subservience to empiric formula, always reacts into the character of the empiric; he becomes hour by hour more and more a mere routinist and less and less an artist. Even that which he has once learned and applied with some interest and intelligence tends to become more and more mechanical, and its application more and more an unintelligent and unemotional procedure. It is never brightened and quickened by adaptation to new ends. The machine teacher, like the empiric in every profession, thus becomes a stupefying and corrupting influence in his surroundings; he himself becomes a mere tradesman, and makes his school a mere machine shop.

3. The Third Resource is *authoritative instruction in methods and devices.* At present, the real opposition is not between native skill and experience on the one side, and psychological methods on the other; it is rather between devices picked up no one knows how, methods inherited from a crude past, or else invented, *ad hoc,* by educational quackery—and methods which can be rationally justified—devices which are the natural fruit of knowing the mind's powers and the ways in which it works and grows in assimilating its proper nutriment. The mere fact that there are so many methods current, and constantly pressed upon the teacher as the acme of the educational experience of the past, or as the latest and best discovery in pedagogy, makes an absolute demand for some standard by which they may be tested. Only knowledge of the principles upon which all methods are based can free the teacher from dependence upon the educational nostrums which are recommended like patent medicines, as panaceas for all educational ills. If a teacher is one fairly initiated into the real workings of the mind, if he realizes its normal aims and methods, false devices and schemes can have no attraction for him; he will not swallow them "as silly people swallow empirics' pills"; he will reject them as if by instinct. All new suggestions, new methods, he will submit to the infallible test of science; and those which will further his work he can adopt and rationally apply, seeing clearly their place and bearings, and the conditions under which they can be most effectively employed. The difference be-

tween being overpowered and used by machinery and being able to use the machinery is precisely the difference between methods externally inculcated and methods freely adopted, because of insight into the psychological principles from which they spring.

Summing up, we may say that the teacher requires a sound knowledge of ethical and psychological principles—first, because such knowledge, besides its indirect value as forming logical habits of mind, is necessary to secure the full use of native skill; secondly, because it is necessary in order to attain a perfected experience with the least expenditure of time and energy; and thirdly, in order that the educator may not be at the mercy of every sort of doctrine and device, but may have his own standard by which to test the many methods and expedients constantly urged upon him, selecting those which stand the test and rejecting those which do not, no matter by what authority or influence they may be supported.

II. We may now consider more positively how psychology is to perform this function of developing and directing native skill, making experience rational and hence prolific of the best results, and providing a criterion for suggested devices.

Education has two main phases which are never separated from each other, but which it is convenient to distinguish. One is concerned with the organization and workings of the school as part of an organic whole; the other, with the adaptation of this school structure to the individual pupil. This difference may be illustrated by the difference in the attitude of the school board or minister of education or superintendent, whether state, county, or local, to the school, and that of the individual teacher within the school. The former (the administrators of an organized system) are concerned more with the constitution of the school as a whole; their survey takes in a wide field, extending in some cases from the kindergarten to the university throughout an entire country, in other cases from the primary school to the high or academic school in a given town or city. Their chief business is with the organization and management of the school, or system of schools, upon certain general principles. What shall be the end and means of the entire institution? What subjects shall be studied? At what stage shall they be introduced, and in what sequence shall they follow one another—that is, what shall be the arrangement of the school as to its various parts in time? Again, what shall be the correlation of studies and methods at every period? Shall they be taught as different subjects? in departments? or shall methods be sought which shall work them into an organic whole? All this lies, in a large measure, outside the purview of the individual teacher; once within the institution he finds its purpose, its general lines of work, its constitutional structure, as it were, fixed for him. An individual may choose to live in France, or Great Britain, or the United States, or Canada; but after he has made his choice, the general conditions under which he shall exercise his citizenship are decided for him. So it is, in the main, with the individual teacher.

But the citizen who lives within a given system of institutions and laws finds himself constantly called upon to act. He must adjust his interests and activities to those of others in the same country. There is, at the same time, scope for purely individual selection and application of means to ends, for unfettered action of strong personality, as well as opportunity and stimulus for the free play and realization of individual equipment and acquisition. The better the constitution, the system which he can not directly control, the wider and freer and more potent will be this sphere of individual action. Now, the individual teacher finds his duties within the school as an entire institution; he has to adapt this organism, the subjects taught, the modes of discipline, etc., to the individual pupil. Apart from this personal adaptation on the part of the individual teacher, and the personal assimilation on the part of the individual pupil, the general arrangement of the school is purely meaningless; it has its object and its justification in this individual realm. Geography, arithmetic, literature, etc., may be provided in the curriculum, and their order, both of sequence and coexistence, laid down; but this is all dead and formal until it comes to the intelligence and character of the individual pupil, and the individual teacher is *the medium through which it comes.*

Now, the bearing of this upon the point in hand is that psychology and ethics have to subserve these two functions. These functions, as already intimated, can not be separated from each other; they are simply the general and the individual aspects of school life: but for purposes of study, it is convenient and even important to distinguish them. We may consider psychology and ethics from the standpoint of the light they throw upon the organization of the school as a whole—its end, its chief methods, the order and correlation of studies—and we may consider them from the standpoint of the service they can perform for the individual teacher in qualifying him to use the prescribed studies and methods intelligently and efficiently, the insight they can give him into the workings of the individual mind, and the relation of any given subject to that mind.

Next to positive doctrinal error within the pedagogy itself, it may be said that the chief reason why so much of current pedagogy has been either practically useless or even practically harmful is the failure to distinguish these two functions of psychology. Considerations, principles, and maxims that derive their meaning, so far as they have any meaning, from their reference to the organization of the whole institution, have been presented as if somehow the individual teacher might derive from them specific information and direction as to how to teach particular subjects to particular pupils; on the other hand, methods that have their value (if any) as simple suggestions to the individual teacher as to how to accomplish temporary ends at a particular time have been presented as if they were eternal and universal laws of educational polity. As a result the teacher is confused; he finds himself expected to draw particular practical conclusions from very vague and theoretical educational maxims (e. g., proceed from the whole to the part, from the concrete to the abstract), or he finds himself expected to adopt as rational principles what are mere temporary expedients. It is, indeed, advisable that the teacher should understand, and even be able to criticise, the general principles upon which the whole educational system is formed and administered.

He is not like a private soldier in an army, expected merely to obey, or like a cog in a wheel, expected merely to respond to and transmit external energy; he must be an intelligent medium of action. But only confusion can result from trying to get principles or devices to do what they are not intended to do—to adapt them to purposes for which they have no fitness.

In other words, the existing evils in pedagogy, the prevalence of merely vague principles upon one side and of altogether too specific and detailed methods (expedients) upon the other, are really due to failure to ask what psychology is called upon to do, and upon failure to present it in such a form as will give it undoubted value in practical applications.

III. This brings us to the positive question: In what forms can psychology best do the work which it ought to do?

1. *The Psychical Functions Mature in a certain Order.*—When development is normal the appearance of a certain impulse or instinct, the ripening of a certain interest, always prepares the way for another. A child spends the first six months of his life in learning a few simple adjustments; his instincts to reach, to see, to sit erect assert themselves, and are worked out. These at once become tools for further activities; the child has now to use these acquired powers as means for further acquisitions. Being able, in a rough way, to control the eye, the arm, the hand, and the body in certain positions in relation to one another, he now inspects, touches, handles, throws what comes within reach; and thus getting a certain amount of physical control, he builds up for himself a simple world of objects.

But his instinctive bodily control goes on asserting itself; he continues to gain in ability to balance himself, to co-ordinate, and thus control, the movements of his body. He learns to manage the body, not only at rest, but also in motion—to creep and to walk. Thus he gets a further means of growth; he extends his acquaintance with things, daily widening his little world. He also, through moving about, goes from one thing to another—that is, makes simple and crude *connections* of objects, which become the basis of subsequent relating and generalizing activities. This carries the child to the age of twelve or fifteen months. Then another instinct, already in occasional operation, ripens and takes the lead—that of imitation. In other words, there is now the attempt to adjust the activities which the child has already mastered to the activities which he sees exercised by others. He now endeavours to make the simple movements of hand, of vocal organs, etc., already in his possession the instruments of reproducing what his eye and his ear report to him of the world about him. Thus he learns to talk and to repeat many of the simple acts of others. This period lasts (roughly) till about the thirtieth month. These attainments, in turn, become the instruments of others. The child has now control of all his organs, motor and sensory. The next step, therefore, is to relate these activities to one another consciously, and not simply unconsciously as he has hitherto done. When, for example, he sees a block, he now sees in it the possibility of a whole series of activities, of throwing, building a house, etc. The head of a broken doll is no longer to him the mere thing directly before his senses. It symbolizes "some fragment of his dream of human life." It arouses in consciousness an entire group of related actions; the child strokes it, talks to it, sings it to sleep, treats it, in a word, as if it were the perfect doll. When this stage is reached, that of ability to see in a partial activity or in a single perception, a whole system or circuit of relevant actions and qualities, the imagination is in active operation; the period of symbolism, of recognition of meaning, of significance, has dawned.

But the same general process continues. Each function as it matures, and is vigorously exercised, prepares the way for a more comprehensive and a deeper conscious activity. All education consists in seizing upon the dawning activity and in presenting the material, the conditions, for promoting its best growth—in making it work freely and fully towards its proper end. Now, even in the first stages, the wise foresight and direction of the parent accomplish much, far more indeed than most parents are ever conscious of; yet the activities at this stage are so simple and so imperious that, given any chance at all, they work themselves out in some fashion or other. But when the stage of conscious recognition of meaning, of conscious direction of action, is reached, the process of development is much more complicated; many more possibilities are opened to the parent and the teacher, and so the demand for proper conditions and direction becomes indefinitely greater. Unless the right conditions and direction are supplied, the activities do not freely express themselves; the weaker are thwarted and die out; among the stronger an unhappy conflict wages and results in abnormal growth; some one impulse, naturally stronger than others, asserts itself out of all proportion, and the person "runs wild," becomes wilful, capricious, irresponsible in action, and unbalanced and irregular in his intellectual operations.

Only knowledge of the order and connection of the stages in the development of the psychical functions can, negatively, guard against these evils, or, positively, insure the full meaning and free, yet orderly or law-abiding, exercise of the psychical powers. In a word, education itself is precisely the work of supplying the conditions which will enable the psychical functions, as they successively arise, to mature and pass into higher functions in the freest and fullest manner, and this result can be secured only by knowledge of the process—that is, only by a knowledge of psychology.

The so-called psychology, or pedagogical psychology, which fails to give this insight, evidently fails of its value for educational purposes. This failure is apt to occur for one or the other of two reasons: either because the psychology is too vague and general, not bearing directly upon the actual evolution of psychical life, or because, at the other extreme, it gives a mass of crude, particular, undigested facts, with no indicated bearing or interpretations:

1. The psychology based upon a doctrine of "faculties" of the soul is a typical representative of the first sort, and educational applications based upon it are necessarily mechanical and formal; they are generally but plausible abstractions, having little or no direct application to the practical work of the classroom. The mind having been considered as split up into a number of independent powers, pedagogy is reduced to a set

of precepts about the "cultivation" of these powers. These precepts are useless, in the first place, because the teacher is confronted not with abstract faculties, but with living individuals. Even when the psychology teaches that there is a unity binding together the various faculties, and that they are not really separate, this unity is presented in a purely external way. It is not shown in what way the various so-called faculties are the expressions of one and the same fundamental process. But, in the second place, this "faculty" psychology is not merely negatively useless for the educator, it is positively false, and therefore harmful in its effects. The psychical reality is that continuous growth, that unfolding of a single functional principle already referred to. While perception, memory, imagination and judgment are not present in complete form from the first, one psychical activity is present, which, as it becomes more developed and complex, manifests itself in these processes as stages of its growth. What the educator requires, therefore, is not vague information about these mental powers in general, but a clear knowledge of the underlying single activity and of the conditions under which it differentiates into these powers.

2. The value for educational purposes of the mere presentation of unrelated facts, of anecdotes of child life, or even of particular investigations into certain details, may be greatly exaggerated. A great deal of material which, even if intelligently collected, is simply data for the scientific specialist, is often presented as if educational practice could be guided by it. Only *interpreted* material, that which reveals general principles or suggests the lines of growth to which the educator

has to adapt himself, can be of much practical avail; and the interpreter of the facts of the child mind must begin with knowing the facts of the adult mind. Equally in mental evolution as in physical, nature makes no leaps. "The child is father of the man" is the poetic statement of a psychologic fact.

3. *Every Subject has its own Psychological Place and Method.*—Every special subject, geography, for instance, represents a certain grouping of facts, classified on the basis of *the mind'sattitude towards these facts.* In the thing itself, in the actual world, there is organic unity; there is no division in the facts of geology, geography, zoölogy, and botany. These facts are not externally sorted out into different compartments. They are all bound up together; the facts are many, but the thing is *one.* It is simply some interest, some urgent need of man's activities, which discriminates the facts and unifies them under different heads. *Unless the fundamental interest and purpose which underlie this classification are discovered and appealed to, the subject which deals with it can not be presented along the lines of least resistance and in the most fruitful way.* This discovery is the work of psychology. In geography, for example, we deal with certain classes of facts, not merely in themselves, but from the standpoint of their influence in the development and modification of human activities. A mountain range or a river, treated simply as mountain range or river, gives us geology; treated in relation to the distribution of genera of plants and animals, it has a biological interpretation; treated as furnishing conditions which have entered into and modified human activities—grazing,

transportation of commodities, fixing political boundaries, etc.—it acquires a geographical significance.

In other words, the unity of geography is a certain unity of human action, a certain human interest. Unless, therefore, geographical data are presented in such a way as to appeal to this interest, the method of teaching geography is uncertain, vacillating, confusing; it throws the movement of the child's mind into lines of great, rather than of least, resistance, and leaves him with a mass of disconnected facts and a feeling of unreality in presence of which his interest dies out. All method means adaptation of means to a certain end; if the end is not grasped, there is no rational principle for the selection of means; the method is haphazard and empirical—a chance selection from a bundle of expedients. But the elaboration of this interest, the discovery of the concrete ways in which the mind realizes it, is unquestionably the province of psychology. There are certain definite modes in which the mind images to itself the relation of environment and human activity in production and exchange; there is a certain order of growth in this imagery; to know this is psychology; and, once more, to know this is to be able to direct the teaching of geography rationally and fruitfully, and to secure the best results, both in culture and discipline, that can be had from the study of the subject.

Application to Arithmetic.—In the following pages an attempt has been made to present the psychology of number from this point of view. Number represents a certain interest, a certain psychical demand; it is not a bare property of facts, but is a certain way of interpreting and arranging them—a certain method of construing them. What is the interest, the demand, which gives rise to the psychical activity by which objects are taken as numbered or measured? And how does this activity develop? In so far as we can answer these questions we have a sure guide to methods of instruction in dealing with number. We have a positive basis for testing and criticising various proposed methods and devices; we have only to ask whether they are true to this specific activity, whether they build upon it and further it. In addition to this we have a standard at our disposal for setting forth correct methods; we have but to translate the theory of mental activity in this direction—the psychical nature of number and the problem of its origin—over into its practical meaning. Knowing the nature and origin of number and numerical properties as psychological facts, the teacher knows how the mind works in the construction of number, and is prepared to help the child to think number; is prepared to use a method, helpful to the normal movement of the mind. In other words, rational method in arithmetic must be based on the psychology of number.

CHAPTER II.

THE PSYCHICAL NATURE OF NUMBER.

Why do we ask with respect to any magnitude, "how many," "how much," and set about counting and measuring till we can say "so many," "so much"? Why do we not take our sense experience just as it comes to us, making no attempt to give it these exact quantitative characteristics? If we can find out the psychological reason, the mental necessity, which induces us to put our experience so far as we can into terms of exact measurement, we shall have a principle which will guide us to sound conclusions regarding the nature and origin of number and its rational treatment as a school study. We have here tacitly assumed that number is a psychical product, and has a psychical reason for its origin. Before dealing with the problem of the origin of number, let us put the assumption of its psychical nature on a firmer basis.

Number is a Rational Process, not a Sense Fact.—The mere fact that there is a multiplicity of things in existence, or that this multiplicity is present to the eye and ear, does not account for a consciousness of number. There are hundreds of leaves on the tree in which the bird builds its nest, but it does not follow that the bird can count.

So hundreds of noises strike the ear, and countless objects appeal to the eye of a child a few weeks old; but he is not conscious of the noises or the objects as quantitative; he does not number or measure them. More than this, sense facts may be even attended to without giving the idea of number. To put, say, five objects before an older child, to call his mind away from all other things and get his attention fixed upon these objects, is not to give him the idea of the number five. Number is not a property of the objects which can be realized through the mere use of the senses, or impressed upon the mind by so-called external energies or attributes. Objects (and measured things) aid the mind in its work of constructing numerical ideas, but the objects are not number. Nor does the bare perception of them constitute number. A child, or an adult, may perceive a collection of balls or cubes, or dots on paper, or a bunch of bananas, or a pile of silver coins, without an idea of their number; there may be clear and adequate percepts of the things quite unaccompanied by definite numerical concepts. No such concepts, no clearly defined numerical ideas, can enter into consciousness till the mind orders the objects—that is, compares and relates them in a certain way.

FACTORS OF THE INTELLECTUAL PROCESS.—In the simple recognition, for example, of three things as three the following intellectual operations are involved: *The recognition of the three objects as forming one connected whole or group*—that is, there must be a recognition of the *three* things as individuals, and of the *one*, the unity, the whole, made up of the three things. If one of the objects is a piece of candy, and the other two are dots on paper, the candy may so absorb attention that the two dots do not present themselves in consciousness at all. This is undoubtedly one reason why the mathematical attainments of savages are so meagre; they are so given up to one absorbing thing—which is to them what the candy is to the child—that the rest of the universe, however much it may affect their senses, does not become an object of attention.

Or, again, the child may be conscious of the dots as well as of the candy, and yet not be able to recognise that these various objects are connected or make one whole. The qualitative unlikeness of the objects may be so great as to make it difficult or even impossible for the child's mind to relate them, to view them all from a common standpoint as forming one group. The candy is one thing and the dots are another and entirely different thing. Here, again, rational counting is out of the question.

Nor, finally, is it to be concluded that from the mere presentation of three *like* objects the idea of three will be secured. There must be enough qualitative unlikeness—if only of position in space or sequence in time—to mark off the individual objects, to keep them from fusing or running into one vague whole. Part of the difficulty of performing the abstraction which is required to get the idea of number is, accordingly, that this abstraction is *complex*, involving two factors: the difference which makes the individuality of each object must be noted, and yet the different individuals must be grasped as one whole—a *sum*. It requires, then, considerable power of intellectual abstraction even to count three. Unlike objects, in spite of differences in quality, must be recognised as forming *one* group; while a group of like objects, in spite of their similarities in quality, is to be recognised as made up of separate things. *Three* differently coloured cubes, for example, must be apprehended as *one* group, while a group of three cubes exactly alike must be apprehended as *three* individuals. In other words, the objects counted, whatever be their physical resemblances or differences, are numerically alike in this: *they are parts of one whole*—they are *units* constituting a defined *unity*.

The delight which a child four or five years old often manifests in the apparently mechanical operation of counting chairs, books, slate-marks, playthings, or even in simply saying over the names of number symbols is really delight in his newly acquired or rapidly growing power of abstraction and generalization. There is abstraction because the child now knows, in a definite, objective way, that one chair, although a *different* chair from every other, is, nevertheless, in some particular identical with every other—it is a chair. He is able to neglect all that sensuous qualitative difference which previously so claimed his attention as to prevent his conscious or objective recognition of the common quality or use through which the things may be classed as one whole. Now, ability to neglect certain features of things in view of another considered more important, is of course of the essence of abstraction in its highest as well as in this rudimentary form. Generalization, on the other hand, is simply the obverse of abstraction; they are correlative phases of one activity. In leaving out of account the qualities now seen to be unimportant to the end in view, though sensuously they may be very prominent and attractive, the mind grasps in one whole the objects that have a *common* quality or use, though the objects are decidedly unlike as regards other qualities or uses. If from a collection of objects of different colours a child is required to select all the *red* ones, he not only neglects all that are not red; he neglects also all the other qualities—shape, size, material, etc.—of the red objects themselves; and when this *abstraction* is completed, there is the conception of the group of red things as the result of the other side of the mental process—viz., *generalization*.

The manifestation of the conscious tendency in a child to count coincides, then, with the awakening in his mind of conscious power to abstract and generalize. This power can show itself only when there is ability to resist the immediate solicitations of colour, sound, etc., ability to hold the mind from being absorbed in the delight of mere seeing, hearing, handling; and this means power of *abstraction*. But this very power to resist the stimulus of some sense qualities and to attend to others means also the power to group the different objects together on the basis of some principle not directly apprehended by the senses—some use or function which all the different objects have—and this is, again, generalization.

DISCRIMINATION AND RELATION.—This power to form a whole out of different objects may be studied in somewhat more detail. It includes the two correlative powers of discrimination and relation.

1. *Discrimination*.—As adults we are constantly deceiving ourselves in regard to the nature and genesis of our mental experiences. Because an object presents

a certain quality directly to us, we are apt to assume that the quality is inherent in the object itself, and is presented to everybody quite apart from any intellectual operation. We forget that the objects *now* have certain qualities for us *simply because of analyses previously performed*. We see in an object just what we have learned to see in it. The contents of the concept resulting from an elaborate process of analysis-synthesis are at last given in the percept. An expert geometrician's percept of a triangle is quite a different thing from that of a mere tyro in geometry. A man may become such a chemist as never to see water without being conscious that it is composed of oxygen and hydrogen; or such a botanist that a passing glance at a flower instantly recalls the name orchid, or ranunculus, and all the differential qualities which belong to this class of plant life. In like manner all of us have become sufficiently familiar with numerical ideas to know at a glance that a tree has a great many leaves, a chair a certain number of parts, a cube a definite number of faces. Although this knowledge is now direct and "intuitive," it is the result of *past* discriminations. We may be perfectly sure that they are not "intuitions" to the child; to him the tree, the house, the cube, the blackboard, the group of six objects, is one undefined whole, not a whole of parts. The recognition of separate or distinct parts always implies an *act of analysis or discrimination definitely performed at some period;* and such definite analysis has always been preceded by a vague synthesis—that is, the idea of a whole of as yet undistinguished parts.

There is perhaps no point at which the teacher is

more likely to go astray than in assuming that objects have for a child the definiteness or concreteness of qualities which they have for us. In the application of the pedagogical maxim "from the concrete to the abstract," he is very apt to overlook the necessity of making sure that the "concrete" is really present to the child's mind. He too easily assumes as already existing in the consciousness of the learner what can really exist only as the product of the mind's own activity in the process of definition—of discriminating and relating. It is a grave error to suppose that a triangle, a circle, a written word, a collection of five objects, are concrete wholes, that is, definitely grasped *mental* wholes to the child, simply because there are certain physical wholes present to his senses. Definite *ideas* are thus assumed as the basis of later work when there is absolutely nothing corresponding to them in the child's mind, in which, indeed, there is only a panorama of vague shifting imagery, with a penumbra of all sorts of irrelevant emotions and ideas. Thus, this noted maxim, when translated to mean *concrete things before the senses, therefore concrete knowledge in the mind*, becomes really a mischievous fallacy.

Or, again, the teacher, mislead by the formula—first, the isolated definite particular; second, the interconnection; third, the organic whole—introduces distinction and definition where normally the child should deal only with wholes in vague outline; and thus substitutes for the poetic and spontaneous character of mental action a forced mechanical analysis all out of harmony with his existing stage of development. Of this we have an example in the prevailing methods of primary number teaching. The child is from the beginning drilled in the "analysis" of numbers till he knows or is supposed to know "all that can be done with numbers." It appears to be forgotten that he may and should perform many operations and reach definite results by implicitly *using* the ideas they involve long before these ideas can be explicitly developed in consciousness. If facts are presented in their proper connection as stimulating and directing the primary mental activities, the child is slowly but surely feeling his way towards a conscious recognition of the nature of the process. This unconscious growth towards a reflective grasp of number relations is seriously retarded by untimely analysis—untimely because it appeals to a power of reflection which is as yet undeveloped.

It is obvious that these two errors are logically opposed to each other. One overlooks the need of the process of discrimination, of careful analysis; the other does nothing but analyze and define. But while logically opposed to each other they are often practically combined. They both arise from one fundamental error—the failure to grasp clearly the place which discrimination occupies as the transitional step in the change of a vague whole into a coherent whole. In the ordinary methods of teaching number, for example, both mistakes are found in combination. There is no attention, or too little attention, paid to the essential process of discrimination when it is taken for granted that definite ideas of number will be formed from the hearing and memorizing of numerical tables, or even from the perception of certain objects *apart from the child's own activity in conceiving a whole of parts and relating parts in a definite whole.* On the other hand, there is altogether too much definition, definition carried to the point of isolation, when, in number teaching, a start is made with one thing—endless changes being rung with single objects in order "to develop the number one"—then another object is introduced, then another, and so on. Here the preliminary activity that resolves a whole into parts is omitted, as well as the connecting link *that makes a whole of all the parts.*

2. *Relation or Rational Counting.*—This involves the putting of units (parts) in a certain ordered relation to one another, as well as marking them off or discriminating them. If, when the child discriminates one thing from another, he loses sight of the identity, the link which connects them, he gains no idea of a group, and hence there is no counting. There is, to him, simply a lot of unrelated things. When we reach "two" in counting, we must still *keep in mind "one"*; if we do not we have not "two," but merely *another* one. Two things may be before us, and the word "two" may be uttered but the concept two is absent. The concept two involves the act of putting together and holding together the two discriminated ones. It is this tension between opposites which is largely the basis of the childish delight in counting. Number is a continued paradox, a continued reconciliation of contradictions. If two things are simply *fused* in each other, forming a sort of vague *oneness*, or if they are simply *kept apart* from each other, there is no counting, no "two." It is the correlative differentiation and identification, the holding apart and at the same time bringing

together, which imparts to the operation of counting its fascination. This activity is simply the normal exercise of what are always the fundamental rational functions; and thus it gives to the child the same sense of power, of ease and mastery in mental movement, that an adult may realize from some magnificent generalization through which a vast, disorderly field of experience is reduced to unity and system. In the simple one, two, three, four of the child, as he counts the familiar objects around him, there is presented the *form* of the highest operations of discrimination and identification.

EDUCATIONAL SUMMARY.—*The idea of number is not impressed upon the mind by objects even when these are presented under the most favourable circumstances. Number is a product of the way in which the mind deals with objects in the operation of making a vague whole definite.* This operation involves (*a*) *discrimination* or the recognition of the objects as distinct individuals (units); (*b*) *generalization*, this latter activity involving two subprocesses; (*1*) *abstraction*, the neglecting of all characteristic qualities save just enough to limit each object as *one*; and (*2*) *grouping*, the gathering together the like objects (units) into a whole or class, the *sum*. Hence:

1. Number can not be taught by the mere presentation of things, but only by such presentation as will stimulate and aid the mental movement of discriminating, abstracting, and grouping which leads to definite numerical ideas.

2. In this process there must be sufficient qualitative difference among the objects used to facilitate the recognition of individuals as distinct, but not enough to resist

the power of grouping all the individuals, of grasping them as parts of one whole or sum.

The application of this principle will depend largely upon circumstances (sensory aptitudes, etc.) and the tact of the teacher. In some cases it may be well at the outset to use differently coloured cubes, the different colours serving to individualize each object or group of objects as a unit, while the common cubical quality facilitates relation. In other cases the difference in colour might divert attention from the relating process, and hinder the grasping of the different *units* as one *sum;* the mere difference of position in space would be enough for the necessary discrimination.

3. In any case the aim must be to enable the pupil to get along with the minimum of actual sense difference, and thus further the power of mathematical abstraction and relation. For discrimination must operate just enough for the recognition of the individuality or singleness of each object or part, and no further. The end is the facile recognition of groups as *groups*, the individuals, the single, component parts being considered not for their own sake, but simply as giving definite value to the group. That is to say, the recognition, for instance, of three, or four, or five, must be as nearly as possible an intuition; a perception of the parts in the whole or a whole of parts, and not a conscious recognition of each part by itself, and then a conscious uniting it to other parts separately recognised.

4. It is clear that to promote the natural action of the mind in constructing number, the starting point should be not a single thing or an unmeasured whole, but a group of things or a measured whole. Attention

fixed upon a single unmeasured object will discriminate and unify the qualities which make the thing a qualitative whole, but can not discriminate and relate the parts which make the thing a definite quantitative whole. It is equally clear that with groups of things the movement in numerical abstracting and relating may be greatly assisted by the arrangement of the things in analytical forms, as is the case, e. g., with the points on dominoes.

CHAPTER III.

THE ORIGIN OF NUMBER: DEPENDENCE OF NUMBER ON MEASUREMENT, AND OF MEASUREMENT ON ADJUSTMENT OF ACTIVITY.

ADMITTING, then, the psychical nature of number, we are now prepared to deal with its psychological origin. It does not arise, as we have seen, from mere sense perception, but from certain rational processes in construing, in defining and relating the material of sense perception. But we are not to suppose that these processes—numerical abstraction and generalization—account for themselves. They give rise to number, but there is some reason why we perform them. This reason we must now discover, for it lies at the root of *the problem of the origin of number*.

THE IDEA OF LIMIT.—If every human being could use at his pleasure all the land he wanted, it is probable that no one would ever measure land with mathematical exactness. There might be, of course—Crusoe-like—a crude estimate of the quantity required for a given purpose; but there would be no definite numerical valuation in acres, rods, yards, feet. There would be no need for such accuracy. If food could be had without trouble or care, and in sufficiency for everybody, we should never put our berries in quart measures, count off eggs and oranges by the dozen, and weigh out flour by the pound. If everything that ministers to human wants could be had by everybody just when wanted, we should never have to concern ourselves about quantity. If everything with which human activity is in any way concerned were unlimited, there would of course be no need to inquire respecting anything whatever: What are its limits? How much is there of it? Even if a thing were not actually unlimited, if there were always enough of it to be had with little or no expenditure of energy, it would be *practically* unlimited, and hence would never be measured. It is because we have to put forth effort, because we have to take trouble to get things, that they are limited for us, and that it becomes worth while to determine their limits, to find out the *quantity* of anything with which human energy has to do.

Limit, in other words, is the primary idea in all quantity; and the idea of limit arises because of some resistance met in the exercise of our activity.

Economy of Energy.—Because we have to put forth effort, because we are confronted by obstacles, our energy is limited. It therefore becomes necessary to economize our energy—that is to say, to dispose of it or distribute it in such ways as will accomplish the best possible results. This economy does not mean a hoarding up or withholding of energy, but rather *giving it out* in the *most effective way,* husbanding "our means so well they shall go far." If we put forth more energy than is needed to effect a certain purpose, and equally if we put forth less than is needed, there is waste; we fail to make the most of the resources at our disposal. We carry out our plans most successfully, and perform the hardest tasks with the least waste of power when we accurately adjust our energies to the thing required. Because of the limitation of human energy all activity is a balancing of energy over against the thing to be done, and is most fruitful of results when the balancing is most accurate. If the arrow of the savage is too heavy for his bow, or if it is too light to pierce the skin of the deer, there is in both cases a waste of energy. If the bow is so thick and clumsy that all his strength is required to bend it, or so slight or uneven that too little momentum is given to the arrow, there is but a barren show of action, and the savage has his labour for his pains. Bow and arrow must be accurately adjusted to each other in size, form, and weight; and both have to be equated (as the mathematician would say) or balanced to the end in view—the killing of the game. This involves the process of measurement, and its result is more or less definite *numerical values.*

Means and End: Valuation.—The same principle may be otherwise stated in terms of the relation existing between means and end. If all our aims were reached at the moment of forming them, without any delay, postponement, or countervening occurrences—if to realize an end we had only to conceive it—the necessity for measurement would not exist, and there would be no such thing as number in the strictly mathematical sense. But the check to our activity, the limitation of energy, defers the satisfaction of our needs. The end to be realized is remote and complex, and in using adequate means, distance in space, remoteness in time, quantity of some sort has to be taken into account, and this means accurate measurement.

In working out a certain purpose, for example, one of a series of means is a journey to be undertaken; it is of a certain length; it is to be completed in a given time, and within a certain maximum of expense, etc.; and this involves careful calculation—measurement and numerical ideas. In brief, it may be said that *quantity* enters into all the activities of life, that the limitation of energy demands its economical use—that is, the precise adjustment of means to end—and that such use, such careful adjustment of activity, depends upon exact *measurement* of quantity.

The child and the savage have very imperfect ideas of number, because they are taken up with the things of the present moment. There is no imperative demand for the economical adjustment of means to end; living only in and for the present, they have no plans and no distant end requiring such an adjustment. They do not ask how the present is to be made of use in attaining some future or permanent good. But as soon as the child or the savage has to arrange his acts in a certain order, to prescribe for himself a certain course of conduct so as to accomplish something remote, then the idea of quantity begins to exist. When a savage is aimlessly playing with a stick, he does not connect it with any desired end, and accordingly does not reflect upon its quantitative value. But if he wants to shape it into an arrow, then the measuring (the quantifying) immediately begins; he examines several sticks; he thinks this one too big, that too small; this too brittle, that too elastic; this one of the right size, weight, and elasticity —all of which are simple quantitative ideas.

Thus it is also in the case of a child playing with

stones; so long as he is contented with them just as they are, not thinking of them as means towards a definite end, ideas as to their weight, or size, or number, do not enter into his mind. But if he decides to throw at a mark, a rude measurement or valuation begins; this is too heavy or too large, that too light or too small, etc. Or if he wishes to build a house, or to form an inclosure with them, he begins at once to note size and shape, and perhaps to form a vague notion of the number required for his ideal house or inclosure. Having only the vaguest ideas of quantity and number, he can not accurately compare means with end, and his first efforts at building will be purely tentative. His ideal "playhouse" or sheepfold, or garden, is too large for his means; he has not stones enough to complete the house or to "go round" the inclosure; he must build on a smaller scale. Or the structures are completed without exhausting the materials, and with the remaining stones he puts together another piece of work, perhaps an addition to house or garden. In all this there is more careful comparison of quantities, a better adjustment of means to end, closer measurement, and some approach to definite numerical ideas.

Again, the savage has to take a journey to find his game. The remoteness of the hunting ground makes him a "measurer"; he must think how *distant* the hunting ground is, how *long* he will be gone, how *much* he should carry with him, etc. If he has a choice of ways of reaching his hunting ground, the numerical valuation becomes more marked: short and long, near and far become more closely defined. The comparison of different means as to their serviceableness in reaching an end not only gives us a vague idea of their quantity, but tends to make it precise, numerical.

The importance of this process of comparing different means in order to select the best, in the development of number judgments, may be illustrated as follows : A chick just out of the shell will peck accurately at a grain of corn or at a fly. We might say that it measured the distance. But this does not mean that it has any *idea* of distance or that there is any conscious process of estimating its extent. There is, in reality, no measuring, no comparison, no selection, but simply direct response to the stimulus; and there is, therefore, no sense of distance. So an average child by the time he is six months old will reach out only for objects which fall within the length of his arms, while previously he may have attempted to grasp objects irrespective of their distance. Yet he does not measure, or necessarily have a consciousness of distance, unless there happen to be, say, two objects, one just within his reach, the other just beyond, and he selects the nearest object as a result of comparing the distances. If the child does this, he performs a rudimentary measurement and has a crude idea of distance. So, too, creeping, walking, etc., imply what may be termed measurement, but they involve no *process* of measuring, and hence no consciousness of space values, of length, or size, or form, until the child begins to prefer and select one of several paths. When he does this, he refers the various paths to his own *ease of action*, and thus gets a standard for comparison.

Summary.—The conscious adjusting of means to

end, particularly such an adjusting as requires comparison of different means to pick out the fittest, is the source of all quantitative ideas—ideas such as more and less, nearer and farther, heavier and lighter, etc. Quantity means the *valuation* of a thing with reference to some end; what is its *worth, its effectiveness*, compared with *other possible means*. These two conceptions— (*a*) the origin of quantitative ideas in the process of valuation (measuring) and (*b*) the dependence of valuation upon the adjusting of means to an end (i. e., ultimately upon activity) are the beginning of all conceptions of quantity and number, and the sound basis of all dealing with them.

THE IDEA OF BALANCE OR EQUATION.—We shall now note more definitely what is implied in the foregoing account—the constant aiming at a balance or equivalence—(*valens*, worth; *æquus*, equal). The process of adjusting means to end is not simply a process of roughly estimating the value of certain things with regard to the end aimed at; but, as already said, it is economical and successful in the degree in which is employed just the amount of energy, just the amount of means necessary to accomplish the aim. The means used must just balance the end sought. Every machine, for example, represents an adjustment of certain means to a certain end. But there are good machines and bad machines. What constitutes the difference between a good machine and a poor one? The difference is found precisely in the fact that the former represents in itself and in the arrangement of its parts not only an adjustment in general to *some* end, but an *accurate* adjustment to the *precise* end to be reached; it is the embodiment in wood and iron of a series of mechanical principles—it represents an equation. Now it is this necessity of exact balance or equivalency which transforms the vague quantitative ideas of smaller and greater, heavy and light, and so on, into the definite quantitative ideas of just so distant, just so long, so heavy, so elastic, etc. This demands the introduction of the idea of *number*. Number is the definite measurement, the definite valuation of a quantity falling within a given limit. Except as we count off means and end into just so many definite units, there can not be an economical adjustment, and there can not be a precise balance.

Summary.—Number arises in the process of the exact measurement of a given quantity with a view to instituting a balance, the need of this balance, or accurate adjustment of means to end, being some limitation.

Illustration.—The logical steps of the development of number may be illustrated as follows: First, there is a recognition that one is distant from one's destination —say, the camp. Next, that one can by travelling fast reach the camp by sunset. Third, the recognition that the present time (the time of starting) is such an hour of the day, e. g., two o'clock, and that sunset occurs at such a time—say seven o'clock; that the present spot is just so many miles distant from the camp; and that, consequently, one will have to travel just so many miles in a given time—say an hour—to reach the camp; this last stage being the equation or balance.

THE REASON FOR ABSTRACTION AND GENERALIZATION.—We are now prepared to see the reason for the neglect of the sense qualities (the abstraction) and for the reference to the whole (the generalization) included in all numbering. When we are regarding a thing not in itself, but simply as a means for some end, we take no account of any qualities which it may possess except this one quality of being related to the end. If I am to find out merely the quantity of land in a field, the fact that a part of the field is heavy clay and the rest rich, loamy soil is not taken into consideration; these qualities do not make the size value of the field, and are nothing to my purpose. I restrict attention entirely to the *mathematical* measurements, which in themselves are necessary and sufficient for the end to be reached— the determination of the absolute area of the field. But if I am to compute the money value of the field, and know that the loamy soil has one value and the clayey soil another, these qualities, having a relation to the end in view, would have to be noted as controlling the measurements; while all other qualities—kind of clay, character of loam, moisture, and dryness—would be neglected as not bearing on the question of the money value of the field.

Similarly, if we want to know the whole amount of cloth in a store, we neglect all special qualities of cloth, and abstract the one quality of being cloth; silks, woollens, linens, cottons, however marked their differences, are alike in possessing this one quality that makes for our present purpose—they are all *cloth*. But if we are required to find the total money value, as in taking an inventory, and the different kinds of cloth have different prices, then we should abstract the special quality of being silk, or linen, or woollen, or cotton, and neglect all other qualities, colours, patterns, etc., which are believed to have no effect upon the price. In other words, it is always the *end in view* which decides what qualities we shall pay attention to and what neglect. We abstract, or select, the special quality that *helps* with reference to this end. The rest, for purposes of measurement, are nothing to us.

It is obvious that it is the same reference to the end to be accomplished that constitutes *generalization*. We regard the various objects selected as having a *relation*, as making up one whole or class, because, no matter what their differences in themselves, they all serve the same end. It is this common service in helping towards one and the same end which binds them together, even if to eye or ear the things are entirely different. It is the reference to the end to be reached that controls both the abstraction and the generalization.

The books composing a library may be of many kinds—primers and dictionaries, novels and poems, printed in all languages, with pages of all sizes, and bindings in endless variety, yet as serving the one purpose of communicating intelligence through written or printed symbols they all fall together, and can be counted as making up one group.

The Process of Measuring.

We have now (1) noted the psychological processes of abstraction and generalization involved in all number, and have (2) traced them to the need of an economical adjustment of means to end which makes necessary the process of measuring from which *number* has its genesis. We have now to note in more detail the nature of this latter process.

Stages of Measurement.—We begin with the vague estimate of bulk, size, weight, etc., and go on to its accurate determination from the indefinite how much to the definite so much. It is the difference between saying that iron is heavy and that so much iron at a given temperature and a given latitude weighs just so much; or between saying that the blackboard is of moderate size and that it contains so many square feet. The development from the crude guess to the exact statement depends upon the selection and recognition of a *unit*, the repetition of which in space or time makes up and thus measures the whole. The savage may begin by saying that his camp is so many suns away. Here his unit is the distance he can travel between sunrise and sunset. He measures by a unit of action, but that unit is itself unmeasured—just how much it is he can not say—or he measures by marking off so many paces. The pace is the unit, and is, relatively, more definite or accurate than the day's journey, but, absolutely, it is unmeasured. Only when the unit itself is accurately defined do we pass from vague quantity to precise numerical value.

Quantities in Different Scales of Measurement.—But the process of measurement may be carried a step further. For accurate measurement the unit itself must be measured with a unit of the same kind of quantity; but the unit may also have a defined relation to a different *kind* of quantity. We are thus enabled to compare quantities lying in *different scales* of measurement. We can not, for example, directly compare weights and volumes, but we may compare them indirectly. If we discover, for instance, that four cubic inches of iron weigh as much as twenty-nine cubic inches of water, four cubic inches of gold as much as seventy-seven cubic inches of water, and two cubic inches of honey as much as three cubic inches of water, we have then the means of comparing cubic inches of iron with pennyweights of gold or with pounds of honey. Thus the different scales of volume and weight are brought together by comparing both with a common standard. Take a case of comparing money values. We can directly compare the cost of three yards of calico with that of nine yards of the same quality; but we can not directly compare—as to cost—lengths of calico of different qualities, or a length of calico with one of silk. But if we know that one yard of calico is worth (is measured by) eight cents, and one yard of silk worth one dollar and sixty cents, we can accurately measure the worth of any quantity of calico in terms of any quantity of silk.

If it were not for this discovery of a unit differing in kind from the quantity to be measured, and yet capable of comparison with it, our exact measurements would always be confined within one and the same scale—time, weight, volume, etc.; we should simply have to guess how much of one scale would equal a given quantity of another.

Moreover, the *measurement within a given scale is imperfect* if we have no means of defining some unit of the scale in terms of a different quantity. We may know how much a pound is in terms of an ounce, an ounce in terms of drachms, but we can never get out of this circle. We can never know how much the ounce, the pound, etc., really is; our measurement can not reach the highest stage of development—what may be called the scientific stage.

There is perfect measurement only when this stage is reached. The pound of the weight scale, for example, is not perfectly defined in terms of weight (ounces, etc.) alone; the pound is more accurately defined when we discover that it is, say, the amount of copper which under certain conditions will displace such and such an amount of water. Only in some such way as this is our unit ultimately defined, and only when the unit of measure is itself perfectly measured can there be perfectly exact or scientific measurement. This measurement of a quantity in terms of quantity unlike in kind, but alike in some one respect,* is the completion of number as the tool of measurement. Beyond this stage, number can not go, but until it has developed to this point it is an imperfect instrument of measurement. There are therefore three stages of measurement:

1. Measuring with an undefined unit, as in measuring length by the unit "pace," apples by the unit apple, etc.

2. Measuring with a unit itself defined by comparison with a unit of same *kind* of quantity—the yard, the pound, the dollar, etc.

3. Measuring with a unit having a definite relation to a quantity of a different kind.

Counting and Measuring.—It has been said that number originates from measurement; that it is a statement of the numerical value of something. But we are accustomed to distinguish counting (i. e., numeration,

---

* This common basis of comparison is always, ultimately, movement in space.

5

numbering) from measuring. It is usually said that we count objects, particular things or qualities, to see how *many* of them there are, while we measure a particular object or quality to see how *much* of it there is. We count chairs, beds, splints, feet, eyes, children, stamens, etc., simply to get their sum total, the *how many;* we measure distance, weight, bulk, price, cost, etc., to see *how much* there is. Some writers say that these "two kinds" of quantity, which they call quantity of magnitude (how much) and quantity of multitude (how many), are entirely distinct. Nevertheless, all counting is measuring, and all measuring is counting. When we count up the number of particular books in a library, we *measure* the library—find out how much it amounts to as a library; when we count the days of the year, we measure the time value of the year; when we count the children in a class, we measure the class as a whole—it is a large or a small class, etc. When we count stamens or pistils, we measure the flower. In short, when we COUNT we *measure.*

On the other hand, in measuring a continuous quantity—"quantity of magnitude"—counting is equally necessary. We may apply a unit of measure to such a quantity and mark off the parts with perfect accuracy, but there is no measurement till we have *counted* the parts. Thus, the only way to measure weight is by counting so *many units* of density; distance, by counting so *many* particular units of length; cost or price, by counting so many units of value—dollars or what not. In other words, when we MEASURE we *count.* The difference is that in what is ordinarily termed counting, as distinct from measuring, we work with an undefined unit; it is vague measurement, because our unit is unmeasured. When we say ten apples, five books, six horses, etc., we measure some whole, some *how much*, by counting its parts, the *how many;* but we do not know just how much one of these parts or units is. If we knew the exact size, or weight, or price of the books or apples, we should have a more accurate measurement and a more accurate valuation.* On the other hand, what we ordinarily call measuring, as distinct from counting, is simply *counting* with a unit which is itself measured by so many definite parts. If I count off four books, "book," the unit which serves as unit of measurement, is itself only a *qualitative*, not a *quantitative* unity, and the quantity four books is not a definitely measured quantity. If I say each book weighs six ounces or is worth sixty cents, the unit of measurement is itself both qualitative and quantitative; and the price or the weight of the four books is a definitely measured quantity.

We shall see hereafter that, strictly speaking, merely qualitative wholes used as units give only addition and subtraction; that the whole which is itself quantitative, as well as qualitative, gives multiplication and division. If, however, the wholes are taken or assumed as equal in value, then, of course, the *operations* of multiplication and division may be performed with them. But this is only because the assumption of equal (measured)

---

* It is a great pity that our authorities use these unmeasured units so much, particularly in fractions. Half an apple, half a pie, is a practical, not a mathematical expression at all. To make it mathematical we should have to know just how great the whole is—how many ounces or cubic inches.

value in the units is made. If we are to divide fifteen apples "equally" among five boys, giving each boy three apples, this "equal" distribution assumes the *equality* of the units (apples) of measure.

Much and Many.—The whole falling within a certain limit supplies the *muchness;* for example, the *amount* of money in a purse, the amount of land in a field, the amount of pressure it takes to move an obstacle, etc. This "much," or amount, is vague and undefined till measured; it is measured by counting it off into so *many* units. We "lay off" distance into so many yards, and then we know it to be so much. We reckon up the pieces of money in the purse and know how much their value is. A man has a pile of lumber; how "much" has he? If the boards are of uniform size, he finds the number (how many) of feet in one board, and counts the number (how many) of boards, and finds the whole *so many* feet—that is, the indefinite "how much" has become, through counting, the definite "so much." Then, again, if he wishes to find the money value of the lumber, how *much* it is worth, he must count off the total number of feet at so much (so many dollars) per thousand, and the resulting so many dollars represents the worth of the lumber. The many, the counting up of the particular units, measures the worth of the whole. The counting has no other meaning, and the measurement of value can occur in no other way.

It is clear that these two sides of all number are relative to each other, just as means and ends are relative. The so many measures the so much, just as the *means* balance the end. The end is the whole, all that comes within a certain limit; the means are the partial activities, the units by which we realize this whole.

To build a house of a certain kind and value, we must have just so many bricks, so many cubic feet of stone, so much lumber, so many days' work, etc. The house is the end, the goal to be reached; these things are the means. The house has been erected at a certain cost; the counting off and valuing of the units which enter into these different factors, is the only way to discover that cost.

CHAPTER IV.

THE ORIGIN OF NUMBER: SUMMARY AND APPLICATIONS.

SUMMARY: *Complete Activity and Subordinate Acts.* —Through the foregoing illustrations—which are illustrations of one and the same principle regarded from different points of view—we are now prepared for the statement which sums up this preliminary examination of quantity. *That which fixes the magnitude or quantity which, in any given case, needs to be measured is some activity or movement, internally continuous, but externally limited. That which measures this whole is some minor or partial activity into which the original continuous activity may be broken up (analysis), and which repeated a certain number of times gives the same result (synthesis) as the original continuous activity.*

This formula, embodying the idea that number is to be traced to measurement, and measurement back to adjustment of activity, is the key to the entire treatment of number as presented in these pages, and the reader should be sure he understands its meaning before going further. In order to test his comprehension of it he may ask himself such questions as these: The year is some unified activity—what activity does it represent? At first sight simply the apparent return of the sun to the same point in the heavens—an external change; yet the only reason for attaching so much importance to this rather than to any other cyclical change, as to make it the unit of time measurement, is that the movement of the sun controls the cycle of human activities—from seedtime to seedtime, from harvest to harvest. This is illustrated historically in the fact that until men reached the agricultural stage, or else a condition of nomadic life in which their movements were controlled by the movement of the sun, they did not take the sun's movement as a measure of time. So, again, the day represents not simply an external change, a recurrent movement in Nature, but a rhythmic cycle of human action. Again, what activity is represented by the pound, by the bushel, by the foot?* What is the connection between the decimal system and the ten fingers of the hands? What activity does the dollar stand for? If the dollar did not represent certain possible activities which it places at our control, would it be a measure of value? Why may a child value a bright penny higher than a dull dollar? And so on.

Illustrations: Stages of Measurement.— Suppose we wish to find the quantity of land in a certain field. The eye runs down the length and along the breadth of the field; there is the sense of a certain amount of movement. This activity, limited by the boundaries of the field, constitutes the original vague muchness—the quantity to be measured—and therefore determines all succeeding processes. Then analysis comes in, the breaking up of this original continuous

---

* The historical origin of these measures will throw light upon the psychological point.

activity into a series of minor, discrete acts. The eye runs down the side of the field and fixes upon a point which appears to mark half the length; this process is repeated with each half and with each quarter, and thus the length is divided roughly into eight parts, each roughly estimated at twenty paces. The breadth of the field is treated in the same way. The eye moves along till it has measured, as nearly as we can judge, just as much space as equals one of the smallest divisions on the other side.

The process is repeated, and we estimate that the breadth contains six of these divisions. Through these interrupted or discrete movements of the eye we are able to form a crude idea of the length and breadth of the field, and thus make a rough estimate of its area. The separate eye movements constitute the analysis which gives the unit of measurement, and the counting of these separate movements (units) is the synthesis giving the total numerical value.

But the breaking up of the original continuous movement into minor units of activity is obviously crude and defective, and hence the resulting synthesis is imperfect and inadequate. The only thing we are certain of is the number of times the minor act has been performed; it is pure assumption that the minor act measures an equal length every time, and a mere guess that each of the lengths is twenty yards. In order, therefore, to make a closer estimate of the content of the field, we may mark off the length and breadth by pacing, and find that it is a hundred and seventy paces in length and a hundred and thirty paces in breadth. This is probably a more correct estimate, because (*a*) we can be much more certain that the various paces are practically equivalent to one another than that the eye movements are equal, and (*b*) since the pace is a more definite and controlled movement, we have a much clearer idea of how much the pace or unit of measurement really is.

But it is still an assumption that the various paces are equal to one another. In other words, this unit of measure is not itself a constant and measured thing, and the required measurement is therefore still imperfect. Hence the substitution for the pace of some measuring unit, say the chain, which is itself defined; the chain is applied, laid down and taken up, a certain number of times to both the length and the breadth of the field. Now the minor act is uniform; it is controlled by the measuring instrument, and hence marks off exactly the same *space* every *time*.* The partial activity being defined, the resulting numerical value—say, eight chains by six chains—is equally definite. Besides, the chain itself may be measured off into a certain number of equal portions; we may apply a minor unit of measure—e. g., the link—until we have determined how many links make up the chain. By means of this analysis into still smaller acts, the meaning of the unit is brought more definitely home to consciousness.†

---

* Note how the two factors of *space* and *time* appear in all measurement, *space* representing concrete value, *time* the abstract number, and both, the measured magnitude.

† If it be noted that all we have done here is to make the original activity of running the eye along length and breadth first continuously and then in an interrupted series of minor movements, more controlled and hence more precise, the meaning of the proposition (page 52) regarding the origin of measurement in the adjustment of

But this mathematical measurement, this analysis-synthesis, is still insufficient for complete adjustment of activity. What, after all, is the value of this measured quality? What is it good for? Until this question is answered there can not be perfect adjustment of activities. To answer this brings us to the third and final stage of number measurement. This field will produce, say, only so many bushels of corn at a given price per bushel; it is, therefore, not worth so much as a smaller field which will produce as much wheat at a larger price per bushel. Or, in addition to the mere size of the field, it may be necessary to take into account not only the value of the crop it will raise, but also the cost of tilling it. Here there must be a much more complete adjustment of activities. The analysis concerns not only so many square rods; it includes also the money value of the crop and the cost of its production. The synthesis compares the result of this complex measurement with the results of other possible distributions of energy. Analytically the conditions are completely defined; synthetically there can be a complete and economical adjustment of the conditions to secure the best possible results.

The measured quantity representing the unified (or continuous) activity is the whole or unity; the measuring parts, representing the minor or partial activities, are the components or units, which make up the unified whole. In all measurement each of these measuring parts in itself is a whole act—as a pace, a day's journey, etc. But in its function of measuring unit it is at once

---

minor acts to constitute a comprehensive activity will be apparent once more.

reduced to a mere means of constructing the more comprehensive act. The end or whole is *one*, and yet made up of *many* parts.

SUMMARY.—All numerical concepts and processes arise in the process of fitting together a number of minor acts in such a way as to constitute a complete and more comprehensive act.

1. This fitting together, or adjusting, or balancing, will be accurate and economical just in the degree in which the minor acts are the same *in kind* as the major. If, for example, one is going to build a stone wall, the use of the means—the minor activities—will not be accurate until one can find a common measure for both the means, the use of the particular stones, and the end, the wall. Size, or amount of space occupied, is this common element. Hence, to define the process in terms of just so many cubic feet required is economical; to describe it in terms of so many stones would be impossible unless one had first found the volumes of the stones. Hence, once more, the abstraction and the generalization involved in all numerical processes—the special qualities of the stone are neglected, and the only thing considered is the number of cubic feet in the stone—abstraction. But through this factor of so much size the stone is referred at once to its place in the whole wall and to the other stones—generalization.

2. An end, or whole of a certain *quality*, furnishes the *limit* within which the magnitude lies. Quantity is limited quality, *and there is no quantity save where there is a certain qualitative whole or limitation.*

3. *Number* arises through the use of means, or

minor units of activity, to construct an activity equal in value to the given magnitude. This process of constructing an equivalent value is *numbering*—evaluation. Hence, there are no *numerical distinctions* (psychologically) except in the process of measuring some qualitative whole.*

4. This measuring or valuation (defining the original vague qualitative whole) will transform the vague quantity into precise *numerical value;* it will accomplish this successfully in just the degree in which the minor activity or unity of construction is itself measured, or is also a numerical value. Unless it is itself a numerical quantity, a unity measured by being counted out into so many parts, the minor and the comprehensive activity can not be made precisely of the same kind. (Principle 1.)

5. Hence the purely correlative character of much and many, of measured whole and measuring part, of value and number, of unity and units, of end and means.

Educational Applications.

We have now to apply the principle concerning the psychological origin of quantity and number to education. We have seen (*a*) the need in life, the demand in actual experience of the race and the individual, which brings the numerical operations; the process of measuring, into existence. We have seen (*b*) what forms number is required to assume in order to meet the need, fulfil the demand. We have now to inquire how far

---

* The pedagogical consequences of neglecting this principle will be seen in discussing the Grube method, or use of the *fixed* unit.

these ideas and principles have a practical application in educational processes.

The school and its operations must be either a natural or an artificial thing. Every one will admit that if it is artificial, if it abandons or distorts the normal processes of gaining and using experience, it is false to its aim and inefficient in its method. The development of number in the schools should therefore follow the principle of its normal psychological development in life. If this normal origin and growth have been correctly described, we have a means for determining the true place of number as a means of education. It will require further development of the idea of number to show the educational principles corresponding to the growth of numerical concepts and operations in themselves, but we already have the principle for deciding how number is to be treated as regards other phases of experience.

THE TWO METHODS: THINGS; SYMBOLS.—The principle corresponding with the psychological law—the translation of the psychological theory into educational practice—may be most clearly brought out by contrasting it with two methods of teaching, opposed to each other, and yet both at variance with normal psychological growth. These two methods consist, the one in teaching number merely as a *set of symbols;* the other in treating it as a *direct property of objects.* The former method, that of symbols, is illustrated in the old-fashioned ways—not yet quite obsolete—of teaching addition, subtraction, etc., as something to be done with "figures," and giving elaborate rules which might guide the *doer* to certain results called "answers."

It is little more than a blind manipulation of number symbols. The child simply takes, for example, the figures 3 and 12, and performs certain "operations" with them, which are dignified by the names addition, subtraction, multiplication, etc.; he knows very little of what the figures signify, and less of the meaning of the operations. The second method, the simple perception or observation method, depends almost wholly upon physical operations with things. Objects of various kinds—beans, shoe-pegs, splints, chairs, blocks—are separated and combined in various ways, and true ideas of number and of numerical operations are supposed necessarily to arise.

Both of these methods are vitiated by the same fundamental psychological error; they do not take account of the fact that number arises in and through *the activity of mind in dealing with objects*. The first method leaves out the objects entirely, or at least makes no reflective and systematic use of them; it lays the emphasis on symbols, never showing clearly what they symbolize, but leaving it to the chances of future experience to put some meaning into empty abstractions. The second method brings in the objects, but so far as it emphasizes the objects to the neglect of the mental activity which uses them, it also makes number meaningless; it subordinates thought (i. e., mathematical abstraction) to things. Practically it may be considered an improvement on the first method, because it is not possible to suppress entirely the activity which uses the things for the realization of some end; but whenever this activity is made incidental and not important, the method comes far short of the intelligence and skill that should be had from instruction based on psychological principles.

While *the method of symbols* is still far too widely used in practice, no educationist defends it; all condemn it. It is not, then, necessary to dwell upon it longer than to point out in the light of the previous discussion *why* it should be condemned. It treats number as an independent entity—as something apart from the mental activity which produces it; the natural genesis and use of number are ignored, and, as a result, the method is mechanical and artificial. It subordinates sense to symbol.

The *method of things*—of observing objects and taking vague percepts for definite numerical concepts—treats number as if it were an inherent property of things in themselves, simply waiting for the mind to grasp it, to "abstract" it from the things. But we have seen that number is in reality a *mode of measuring value*, and that it does not belong to things in themselves, but arises in the economical adaptation of things to some use or purpose. *Number* is not (psychologically) got *from* things, it is put *into* them.

It is then almost equally absurd to attempt to teach numerical ideas and process *without* things, and to teach them simply *by* things. Numerical ideas can be normally acquired, and numerical operations fully mastered only by arrangements of things—that is, by certain acts of mental construction, which are aided, of course, by acts of physical construction; it is not the mere perception of the things which gives us the idea, but the *employing of the things in a constructive way*.

The method of symbols supposes that number arises

wholly as a matter of abstract reasoning; the method of objects supposes that it arises from mere observation by the senses—that it is a property of things, an external energy just waiting for a chance to seize upon consciousness. In reality, it arises from *constructive* (psychical) activity, from the actual use of certain things in reaching a certain end. This method of constructive use unites in itself the principles of both abstract reasoning and of definite sense observation.

If, to help the mental process, small cubical blocks are used to build a large cube with, there is necessarily continual and close observation of the various things in their quantitative aspects; if splints are used to inclose a surface with, the particular splints must be noted. Indeed, this observation is likely to be closer and more accurate than that in which the mere observation is an end in itself. In the latter case there is no interest, no purpose, and attention is laboured and wandering; there is no aim to guide and direct the observation. The observation which goes with constructive activity is a part of the activity; it has all the intensity, the depth of excitation of the activity; it shares in the interest of and is directed by the activity. In the case where the observation is made the whole thing, distinctions have to be *separately* noted and *separately* memorized. There is nothing intrinsic by which to carry the facts noted; that the two blocks here and the two there make four is an *external* fact to be carried by itself in memory. But when the two sets are so used as to construct a whole of a certain value, the fact is *internal*; it is part of the mind's way of acting, of seeing a definite whole through seeing its definite parts. Repetition in one case means simply learning by rote ; in the other case, it means repetition of activity and formation of an intelligent habit.

The rational factor is found in the fact that the constructive activity proceeds upon a principle; the construction follows a certain regular or orderly method. The method of action, the way of combining the means to reach the end, the parts to make the whole, is *relation;* acting according to this relation is rational, and prepares for the definite recognition of reason, for consciously grasping the nature of the operations. Rational action will pass over of itself when the time is ripe into abstract reasoning. The habit of abstracting and generalizing of analysis and synthesis grows into definite control of thinking.

THE FACTORS IN RATIONAL METHOD.—In more detail, dealing with number by itself, as represented by symbols, introduces the child at an early stage to abstractions without showing how they arise, or what they stand for ; and makes clear no reason, no necessity, for the various operations performed, which are all reducible to (*a*) synthesis—addition, multiplication, involution ; and (*b*) analysis—subtraction, division, evolution. The object or observation method shows the relation of number to things, but does not make evident why it has this relation ; does not bring out its value or measuring use, and leaves the operations performed purely external manipulations of number, or rather with things which may be numbered, not internal developments of its measuring power. The method which develops numerical ideas in connection with the construction of some definite thing, brings out clearly (*a*) the natural unity, the limit (the magnitude) to which all number refers; (*b*) the unit of measurement (the particular thing) which helps to construct the whole; and (*c*) the process of measuring, by which the second of these factors is used to make up or define the first—thus determining its numerical value.

(*a*) Only this method presents naturally the idea of a magnitude from which to set out. The end to be reached, the object to be measured, supplies this idea of a given quantity, and thus gives a natural basis for the development and use of ideas of number. In numbers simply as objects, or in things *simply* as observed things, there is no principle of unity, no basis for natural generalization. Only using the various things for a certain end brings them together into one; we count and measure some quantitative *whole*.

(*b*) While every object is a whole in itself, a unity in so far as it represents one single act, no object simply as an observed object is a *unit*. Objects which *we* recognise as three in number may be before the child's senses, and yet there may be no consciousness of them as three different units, or of the sum three. Some writers tell us that each object is one, and so gives the natural basis for the evolution of number; that the starting point is *one* object, to which another object is "added," then a third, etc. But this overlooks the fact that each object is one, not a *unit* but one *whole*, differing from and exclusive of every other whole. That is, to take it as an *observed* object is to centre attention wholly upon the thing itself; attention would discriminate and unify the qualities which make the thing what it is—a *qualitative* whole; but there would be little room for the abstracting and relating action involved in all number. A numerical unit is not merely a whole, a unity in itself, but is, as we have seen, a unity employed as a means for constructing or measuring some larger whole. *Only this use, then, transforms the object from a qualitative unity into a numerical unit.* The sequence therefore is: first the vague unity or whole, then discriminated parts, then the recognition of these parts as measuring the whole, which is *now* a *defined* unity—a sum. Or, briefly, the undefined whole; the parts; the *related* parts (*now* units); the *sum*.

(*c*) Beginning with the numbers in themselves, as represented by mere symbols, or with perceived objects in themselves, there is no intrinsic reason, *no reason in the mind itself*, for performing the operations of putting together parts to make a whole (using the unit to measure the magnitude), or of breaking up a whole into units—discovering the standard of reference for measuring a given unity. These operations,* from either of these standpoints, are purely arbitrary; we may, if we wish, do something with number, or rather with number symbols: the operations are not something that we *must* do from the very nature of number itself. From the point of view of the constructive (or psychical) use of objects, this is reversed. These processes are simply phases of the *act of construction*. Moreover, the operations of addition, multiplication, division, etc., in the method of perceived objects, have to be regarded as

---

* It will be shown in a later chapter that all numerical operations grow out of this fundamental process.

physical heaping up, *physical* increase, *physical* partition ; while in that of number by itself they are purely mental and abstract. From the standpoint of the psychological use of the things, these processes are not performed upon physical things, but with reference to establishing definite values;* while each process is itself concrete and actual. It is not something to be grasped by abstract thought, it is something done.

Finally, to teach symbols instead of number as the instrument of measurement is to cut across all the existing activities, whether impulsive or habitual. To teach number as a property of observed things is to cut it off from all other activities. To teach it through the close adjustment of things to a given end is to re-enforce it by all the deepest activities.

All the deepest instinctive and acquired tendencies are towards the constant use of means to realize ends; this is the law of all action. All that the teaching of number has to do, when based upon the principle of rationally using things, is to make this tendency more definite and accurate. It simply directs and adjusts this process, so that we notice its various factors and measure them in their relation to one another. More-

---

* The complications introduced in schools—e. g., that you can not multiply by a fraction, nor increase a number by division, etc., because multiplication means increase, etc.—result from conceiving the operations as physical aggregation or separation instead of synthesis and analysis of values—mental processes. To multiply $10 by one third is absurd if multiplication means a physical increase; if it means a measurement of value, taking a numerical value of $10 (a measured quantity) in a certain way to find the resulting numerical value, it is perfectly rational.

over, it relies constantly upon the principle of rhythm, the regular breaking up and putting together of minor activities into a whole; a natural principle, and the basis of all easy, graceful, and satisfactory activity.

CHAPTER V.

THE DEFINITION, ASPECTS, AND FACTORS OF NUMERICAL IDEAS.

We may sum up the steps already taken as follows: (1) The limitation of an energy (or quality) transforms it into quantity, giving it a certain undefined muchness or magnitude, as illustrated by size, bulk, weight, etc. (2) This indefinite whole of quantity is transformed into definite *numerical value* through the process of measurement. (3) This measuring takes place through the use of units of magnitude, by putting them together till they make up an equivalent value. (4) Only when this unit of magnitude has been itself measured (has itself a definite numerical value) is the measurement of the whole magnitude or construction of the entire numerical value adequate. Forty feet denotes an adequately measured quantity, because the unit is itself defined; forty eggs denotes an inadequately measured quantity, because the unit of measure is not definite. Were eggs to become worth, say, twenty times as much as they are now worth, they would be weighed out by the pound—that is, inexact measurement would give way to exact measurement. Having before us, then, the psychological process which constitutes measured quantity, we may define number.

DEFINITION OF NUMBER.—The simplest expression of quantity in numerical terms involves two components:

1. *A Standard Unit; a Unit of Reference.*—This is itself a magnitude necessarily of the same kind as the quantity to be measured. Or, as it may be otherwise expressed, the unity of quantity to be measured and the unit of quantity which measures it are *homogeneous* quantities. Thus, inch and foot (measuring unit and measured unity), pound and ton, minute and hour, dime and dollar are pairs of homogeneous quantities.

2. *Numerical Value.*—This expresses *how many* of the standard units make up, or construct, the quantity needing measurement. Examples of numerical value are: the yard of cloth costs *seventeen* cents; the box will hold *thirty-six* cubic inches; the purse contains *eight* ten-dollar pieces. The seventeen, thirty-six, eight represent just *so many* units of measurement, the cent, the cubic inch, the ten-dollar piece: they express the numerical values of the quantities: they are pure *numbers*, the results of a purely mental process. The numerical value alone represents the relative value or ratio of the measured quantity to the unit of measurement. The numerical value and the unit of measurement taken together express the absolute value (or magnitude) of the measured quantity.

In the teaching of arithmetic much confusion arises from the mistake of identifying numerical value with absolute magnitude—that is, *number*, the instrument of measurement with measured quantity. Number is the product of the mere repetition of a unit of measurement; it simply indicates *how many there are;* it is purely abstract, denoting the series of acts by which the mind constructs defined parts into a unified and definite whole. Absolute value (quantity numerically defined) is represented by the application of this *how many* to magnitude, to quantity—that is, to limited quality. To take an example of the confusion referred to: we are told that division is dividing a (1) number into a (2) number of equal (3) numbers. This definition as it stands has absolutely no meaning; there is confusion of *number* with measured *quantity*. Doubtless the definition is intended to mean: division is dividing a certain definite quantity into a number of definite quantities equal to one another. Only in (2), in the definition as quoted, is the term number correctly used; in both (1) and (3) it means a measured magnitude. A measured or numbered quantity may be divided into a number of parts, or taken a number of times; but no number can be multiplied or divided into parts. Number *simply* as number always signifies how many times one "so much," the unit of measurement, is taken to make up another "so much," the magnitude to be measured. It is, as already said, due to the fundamental activities of mind, discrimination, and relation, working upon a qualitative whole; and we might as well talk of multiplying hardness and redness, or of dividing them into hard and red things, as to talk of multiplying a number or of dividing it into parts.

It may be observed that the problems constantly used in our arithmetics, multiply 2 by 4, divide 8 by 4, are legitimate enough provided they are properly interpreted, if not orally at least mentally, but taken literally are absurd. The first expression means, of course, that a quantity having a value of two units of a certain kind is to be taken four times; and similarly $8 \div 4$ *means* that a total quantity of a certain kind is measured by four units or by two units of the same kind. Of course, in all mathematical calculations we ultimately operate with pure symbols, and the operations do not affect the unit of measure; but in the beginning we should make constant reference to measured quantity, and always should be prepared to interpret the symbols and the processes.

3. Number, then, as distinct from the magnitude which is the unit of reference, and from the magnitude which is the unity or limited quality to be measured, is:

The repetition of a certain magnitude used as the unit of measurement to equal or express the comparative value of a magnitude of the same kind. It always answers the question "How many?"

This "how many" may assume two related aspects: either how many times one part as unit has to be taken or repeated to make up the whole quantity; or how many parts as units, each taken once, compose the whole. In the first case, the times of repetition of the measuring unit is mentally the more prominent; in the second, the actual number of measuring parts; e. g., in thinking of forty yards, we may at one time dwell on the forty *times* the unit is repeated; at another time, on the actual forty parts making the unified *whole*.

As already said, the number and the measuring unit together give the absolute magnitude of the quantity. The number by itself indicates its *relative* value. It *always* expresses ratio*—i. e., the relation which the

---

* Hence, again, the absurdity of multiplying pure number or dividing it into parts. We may divide a ratio, but not into parts.

magnitude to be measured bears to the unit of reference. Seven, as pure number, expresses equally the ratio of 1 foot to 7 feet, of 1 inch to 7 inches, of 1 day to 1 week, of $1,000 to $7,000, and so on indefinitely. Simply as seven it has no meaning, no definite value at all; it only states a possible measurement.

This definition arrived at from psychological analysis is that given by some of the greatest mathematicians on a strictly mathematical basis, as may be seen from comparison with the following definitions:

Newton's.—Number is the abstract ratio of one quantity to another quantity of the same kind.

Euler's.—Number is the ratio of one quantity to another quantity taken as unit.*

PHASES OF NUMBER.—The aspects of number follow directly from what has been said. Quantity, the unity measured, whether a "collection of objects" or a physical whole, is *continuous,* an undefined how *much;* number as measuring value is discrete, how *many.* The magnitude, muchness, *before* measurement is mere unity; *after* measurement it is a sum taken as an integer—that is, an aggregation of parts (units) making up one whole; number as showing how many refers to the units, which put together make the sum. Quantity, measured magnitude, is always concrete; it is a certain kind of magnitude, length, volume, weight, area, amount of cost,

---

* J. C. Glashan, one of the acutest of living mathematicians, defines thus: "A *unit* is any standard of reference employed in counting any collection of objects, or in measuring any magnitude. A number is that which is applied to a unit to express the comparative magnitude of a quantity of the same kind as the unit." (See his Arithmetic for High Schools, etc.)

etc. "Number," as simply defining the how many units of measurement, is always abstract.

The conception of measuring *parts* and of *times* of repetition is inseparable from number as expressing the numerical value of a quantity: as discrete, it is so many parts taken one time—constituting the unity; as abstract, it is one part taken so many times. In the one case, as before suggested, attention is more upon the numbered *parts*, in the other, more upon the *number* of the parts. They are absolutely correlative conceptions of the same measured magnitude. That is, a value of $50 may be regarded as determined by taking $1 *fifty* times, or by taking $50—that is, a *whole* of *fifty parts* —*one* time. The numerical process and the resulting numerical value are the same, however we arrive at the *number*—i. e., the ratio of measured quantity to measuring unit. As this conception of the relation between parts and times in the measurement of quantity is essential to the interpretation of numerical operations, we may give it a little further consideration.

We wish to know the amount of money in a roll of dollar bills. We take five dollars, say, as a convenient measuring unit; we separate our undefined whole into groups of five dollars each; we count these groups and find that there are ten of them—i. e., the numerical value is ten; we have now a definite idea both of the measuring unit and of the times it is repeated, and so have reached a definite idea of the amount of money in the roll of bills. We began with a vague whole, an undefined unity; we broke it up into parts (analysis), and by relating (counting) the parts we arrived at our unity again; the same unity, yet not the same as regards the attitude of the mind towards it. It is now a definite unity constituted by a known number of definite parts; it is a *sum* of *units*. On the analytic side of this defining process the emphasis is on the parts, the units; on the synthetic side the emphasis is on the defined unity, the sum. The parts are means to an end; they exist only for the sake of the end, the sum. The ten units—that is, the unit repeated *ten* times—make up, *are*, the *one* sum—i. e., the sum taken one time.

Further, since the unit of measure is itself measured by a smaller unit, the dollar, the same psychological explanation applies to the measurement of the quantity by means of this smaller unit. The five-dollar unit taken *ten* times is identical with *ten* of these units taken once. We are conscious, also, that any part of this five-dollar unit taken ten times is identical with ten such parts taken once. That is, $1, taken ten times, is a whole of $10 taken once; and since this is true of every dollar in the five, our measurement gives a whole of $10 taken once, a whole of $10 taken twice, and so on; that is, altogether, a whole of $10 taken *five* times. In other words, the measurement, ten groups (or units) of five dollars each necessarily implies the correlative measurement, *five* groups of *ten* dollars each.

This rhythmic process of parting and wholing which leads to all definite quantitative ideas, and involves the correlation of times and parts, may be illustrated by simple intuitions. In measuring a certain length we find it, let us suppose, to contain four parts of three feet each; then the relation between parts (measuring units) and numerical value (times of repetition) may be perceived in the following, where the dots symbolize both times and units of quantity:

```
a    •  •  •  •
b    •  •  •  •
c    •  •  •  •
```

Measuring by the 3-feet unit we count it off four times—that is, the quantity is expressed by 3 feet taken four times. This is represented by the four vertical columns of three minor units each. But this measuring process necessarily involves the correlated process which is expressed by 4 feet taken three times. For, in measuring by *three* feet, and finding that it is repeated four times, we perceive that each of its three parts is repeated *four* times, giving the three horizontal rows *a, b, c*—that is to say, *a* is one whole of 4 feet, *b* a second whole of 4 feet, and *c* a third whole of 4 feet; or, in all, 4 feet taken *three times*. Briefly, 1 foot *four* times is one whole of 4 feet; this is true of every foot of the original measure, 3 feet; and therefore 3 feet *four* times is 4 feet *three* times.

It is clear that the two questions, (*a*) in 12 feet how many counts of 4 feet each, and (*b*) how many feet in each of 4 counts making 12 feet, are solved in *exactly the same way;* neither the *three* counts (times) in the first case nor the *three* feet in the second case can be found *without counting the twelve feet off in groups of four feet each.*

This necessary correlation, in the measurement of quantity, between "parts" and "times"—numerical value of the measuring unit and numerical value of the measured quantity—gives the psychology of the fundamental principle in multiplication known as the law of commutation : the product of factors is the same in whatever order they may be taken—i. e., in the case of two factors, for example, either may be multiplicand or multiplier; *a* times *b* is identical with *b* times *a*.

It is asserted by some writers that this commutative law does not hold when the multiplicand is concrete; "for," we are told, "though there is meaning in requiring $4 to be taken three times, there is no sense in proposing that the number 3 be taken four-dollars times"—which is perfectly true. Nevertheless, the objection seems to be founded on a misconception of the psychical nature of number and the psychological basis of the law of commutation. Psychologically speaking, can the multiplicand *ever* be a pure number? If the foregoing account of the nature of number is correct, the multiplicand, however written, must always be understood to express measured quantity; it is always concrete. As already said, 4 × 3 must mean 4 units of measurement taken three times. If number in itself is purely mental, a result of the mind's fundamental process of analysis-synthesis—what is the meaning of 3 × 4 where both symbols represent pure numbers, and where, it is said, the law of commutation does hold? There is no sense, indeed, in proposing to multiply three by four dollars; but equally meaningless is the proposition to multiply one pure number by another—to take an abstraction a number of times.

Thus, if the commutative law "does not hold when the multiplicand is concrete"—indicating a measured quantity—it does not hold at all; there is no such law. But if the psychological explanation of number as arising from measurement is true, there is a law of commutation. We measure, for example, a quantity of 20 pounds weight by a 4-pound weight, and the result is expressed by 4 pounds × 5, but the psychological correlate is 5 pounds × 4. Here we have true commutation of the factors, inasmuch as there is an interchange of both character and function; the symbol which denotes measured quantity in the one expression denotes pure number in the other, and *vice versa*. If the 4 pounds in the one expression remained 4 pounds in the commuted expression, would there be commutation?

We have referred to the fallacy of identifying actual measuring parts with numerical value; it may now be said that, on the other hand, failure to note their necessary connection—their law of commutation—is often a source of perplexity. To say nothing at present of the mystery of "Division," witness the discussions upon the rules for the reduction of compound quantities and of mixed numbers to fractions. To reduce 41 yards to feet we are, according to some of the rules, to multiply 41 by 3. According to others, this is wrong, giving 123 yards for product; and we ought to multiply 3 feet by 41, thus getting the true result, 123 feet. Some rule-makers tell us that though the former rule is wrong it may be followed, because it always brings the same numerical result as the correct rule, and in practice is generally more convenient. It seems curious that the rule should be always wrong yet always bring the right results. With the relation between parts and times before us the difficulty vanishes. The expression 41 yards denotes a measured quantity; 41 expresses the numerical value of it, and one yard the measuring unit; our conception of the quantity is therefore, primarily, 41 parts of 3 feet each, and we multiply 3 feet by 41; but this conception involves its correlate, 3 parts of 41 feet each; and so, if it is more convenient, we may multiply 41 feet by 3.

A similar explanation is applicable to the reduction, e. g., of $3¾ to an improper fraction. The denominator of the fraction indicates what is, in this case, the direct unit of measure, one of the four equal parts of the dollar; and so we conceive the $3 as denoting 3 parts of 4 units (quarter dollars) each, and multiply 4 by 3; or, as denoting 4 parts of 3 units each, and multiply 3 by 4.

Educational Applications.

1. Every numerical operation involves three factors, and can be naturally and completely apprehended only when those three factors are introduced. This does not mean that they must be always formulated. On the contrary, the formulation, at the outset, would be confusing; it would be too great a tax on attention. But the three factors must be there and must be *used*.

Every problem and operation should (1) proceed upon the *basis* of a total magnitude—a unity having a certain numerical value, should (2) have a certain unit which measures this whole, and should (3) have number—the ratio of one of these to the other. Suppose it is a simple case of addition. John has $2, James $3, Alfred $4. How much have they altogether? (1) The total magnitude, the amount (muchness) altogether, is here the thing sought for. There will be meaning to the problem, then, just in so far as the child feels this amount altogether as the *whole* of the various parts.

(2) The unit of measurement is the one dollar. (3) The number is the measuring of how many of these units there are in all, namely, nine. When discovered it defines or measures the how much of the magnitude which at first is but vaguely conceived. In other words, it must be borne in mind that the *thought of some inclusive magnitude* must, psychologically, precede the operation, if its real meaning is to be apprehended. The conclusion simply defines or states exactly how much is that magnitude which, at the outset, is grasped only vaguely as *mere* magnitude.

Are we never, then, to introduce problems dealing with simple numbers, with numbers not attached to magnitude, not measuring values of some kind; are we not to add 4, 5, 7, 8, etc.? Must it always be 4 apples, or dollars, or feet, or some other concrete magnitude? *No, not necessarily as matter of practice in getting facility in handling numbers.* Number is the tool of measurement, and it requires considerable practice with the tool, as a tool, to handle it with ease and accuracy. But this drill or practice-work in " number " should never be introduced until after work based upon definite magnitudes; it should be introduced only as there is formed the mental habit of continually referring number to the magnitude which it measures. Even in the case of practice, it would be safer for the teacher to call attention to his reference of number to concrete values in every case than to go to the other extreme, and neglect to call attention to its use in defining quantity. For example, when adding " numbers," the teacher might say, " Now, this time we have piles of apples, or we have inches, etc., and we want to see how much we have in all"; or the teacher might ask, at the end of every problem, "What were we counting up or measuring that time?" letting each one interpret as he pleased. Just how far this is carried is a matter of detail; what is not a matter of detail is that the habit of interpretation be formed by continually referring the numbers to some quantity.

2. The unit is never to be taught as a *fixed thing* (e. g., as in the Grube method), but always as a unit of measurement. One is never one thing simply, but always that one thing *used as a basis for counting off and thus measuring some whole or quantity*. Absolutely everything and anything which we attend to is *one*; is made one by the very act of attending. If we could take in the whole system of things in one observation, that would be one; if we could isolate an atom and look at that, it would equally be one. The forest is one when we view it as a whole; the tree, the branch, the stem, the leaf, the cell in the leaf, is equally one when it becomes the object or whole with which we are occupied. But this oneness, this unity possessed by every object of attention, has nothing but the name in common with the numerical unit. In itself it is not quantitative at all; it is mere unity of quality, of meaning. It becomes a quantitative unity (a quantity or magnitude) only when considered as *limited* (page 36), and as an end to be reached by the use of certain means. It becomes a *unit* only when used as one of the means to construct a value equivalent to a certain other value. The assumption that some one object is the natural unit of quantity, which is then increased by bringing in other objects, is the very opposite of the truth; number does

not arise at all until we cease taking objects *as* objects, and regard them simply as parts which make up a whole, as units which measure a magnitude (see pages 24, 42, on Abstraction). It is perfectly clear, therefore, that the method of "*close* observation" of objects is essentially vicious; what is claimed as its merit is in reality a grave defect. The child, according to the advocates of this method, "*sees* what is brought to his notice and sees all about it." But in seeing all about the things there must be neglect of the numerical abstraction which sees nothing about the things save this alone: they are parts of one whole. There may be a discriminating and relating of qualities which give the things individual meaning; but there is not the process—at least the process is impeded—which constructs quantitative units into a defined quantitative whole.

This is plainly so in case of units like dollars, inches, pounds, minutes, etc. They are units not in virtue of any quality absolutely inherent in them, but in virtue of their use in measuring cost, length, weight, duration, etc. It may be said that this is not so in the case of books, apples, boys, etc.; that here each book, apple, etc., is a unit in itself. But this is to fall into the error of separating counting from measuring, already referred to. The book, the crayon, or the cube, is a unity, a whole, in itself, but it is not a unit save as used to count up (value) the total amount. The only point is that this counting gives very crude measurement. The unit, book, pie, and so on, is not itself measured by minor units of the same kind. We are measuring with an unmeasured unit, and so the result of our measurement is exceedingly vague and inaccurate, just as it would be to measure length by steps which had themselves no definite length; cost of other goods by potatoes themselves changing in value, etc.*

Further, since the measuring unit is itself measured, is itself made up of minor or sub-units, it may be of any numerical value, denoting two, three, four, five, six, etc., of such minor units. Twenty-five pounds taken as a basis of measurement is a unit, is *one;* taken with a reference to the minor unit (1 lb.), by which itself is measured, it is a defined unity, a *sum*. Thirty-six feet referred to a measuring quantity of three feet has a numerical value twelve; and the three feet is just as much a unit (one, or 1) as *one* foot is in measuring 12 feet. So the quantities 9 piles of silver coin of ten dollars each, and 25 pages of thirty lines each, have respectively the numerical values of 9 and 25—that is, nine units of measurement ("ones") and twenty-five units ("ones") of measurement. This is the very basis of our system of notation. The numerical value, hundred, applied to any measuring unit, denotes a quantity consisting of 10 ten-units; the number, thousand, measures a quantity which is composed of 10 hundred-units; the number, tenth, applied to any unit, measures that quantity which taken ten times makes up the unit of reference; the number, hundredth, used with any measuring unit denotes that quantity which taken ten times makes up one tenth of the unit of reference, etc.

---

* Hence, once more, the fallacious ideas introduced by our arithmetics in illustrating so much by these unmeasured units partially qualitative and only partially quantitative—the pencil, the apple, the orange, and the universal pie—and so little by the definite units of length, size, weight, money value, etc.

3. The true method, then, may be summarised by saying that the proper introduction to numerical operations is by presenting the material in such a way as to require a mental operation of rhythmic *parting and wholing*—that is, a quality or magnitude is to be presented in such a way as to involve both separation (mental separation, that is, of values, not necessarily physical partition) into parts and the recomposition of the parts into the whole. The analysis gives possession of the unit of measurement; the synthesis, or recomposition, gives the absolute value of the magnitude; the process itself brings out the ratio, the pure number.

We thus see the fundamental fallacy of the Grube method in another light. Just as, upon the whole, it proceeds from the mere observation of objects instead of from the constructive *use* of them, so it works with fixed units instead of with a whole quantity which is measured by the application of a unit of measurement. The superiority of the Grube method to some of the other methods, both in the way of introducing objects instead of dealing merely with numerical symbols, and in the way of systematic and definite instead of haphazard and vague work, has tended to blind educators to its fundamentally bad character, psychologically speaking. There is no need to dwell upon this at length after the previous exposition, but the following points may be noted:

(*a*) In proceeding from one to two, then to three, etc., it leaves out of sight the principle of limit, which is both mathematically and psychologically fundamental. There is no limited quality, no magnitude, with its own intrinsic unity, which sets bounds to and gives the reason for the numerical operation. Number is separated from its reason, its function, measurement of quantity, and so becomes meaningless and mechanical. There is no inner need, no felt necessity, for performing the operations with number. They are artificial. We are dealing with parts which refer to no whole, with units which do not refer to a magnitude. It is as sensible as it would be to make a child learn all the various parts of a machine, and carefully conceal from him the purpose of the machine—what it is for, what it does—and thus make the existence of the parts wholly unintelligible.

(*b*) In beginning with the fixed unit one object (1), then going on to two objects, three objects, then other fixed units, there is no intrinsic psychological connection among the various operations. We *may* add, we *may* subtract, we *may* find a ratio; but addition, subtraction, ratio, remain (psychologically) separate processes. According to true psychology, we begin with a whole of quantity, which on one side is analysed into its units of measurement, while on the other these units are synthesised to constitute the value of the original magnitude; we have parts which refer to a whole, and units which make a sum. Here the addition and subtraction are psychological counterparts; we actually perform both these operations, whether we consciously note more than one of them or not. Similarly, we go through a process of ratio-ing in the rhythmic construction of the whole (much) out of the units (many); the conscious grasp of the principle of ratio will therefore involve no new operation, but simply reflection upon what we have already done. First one process, then another, then another, and so on, is the law of the Grube method—

this, in spite of its maxim to teach all processes simultaneously:* first, a process involving all the numerical operations, then, as the power of attention and interpretation ripens, making the process already performed an object of attention to bring out what is involved, is the psychological law.

4. The method which neglects to recognise number as measurement (or definition of the numerical value of a given magnitude), and considers it simply as a plurality of fixed units, necessarily leads to exhausting and meaningless mechanical drill. The psychological account shows that the natural beginning of number is a whole needing measurement; the Grube method (with many other methods in all but name identical with the Grube) says that some one thing is the natural beginning from which we proceed to two things, then to three things, and so on. Two, three, etc., being fixed, it becomes necessary to master each before going on to the next. Unless four is exhaustively mastered, five can not be understood. The conclusion that six months or a year should be spent in studying numbers from 1 to 5, or from 1 to 10, the learner exhausting all the combinations in each lower number before proceeding to the higher, follows quite logically from the premises. Yet no one can deny that, however much it is sought to add interest to this study (by the introduction of various objects, counting eyes, ears, etc., dividing the children into groups, etc.), the process is essentially one

---

* Which, of course, it never does. It only teaches all of them about one "number" before it goes on to another, each number being an entity in itself—which it ought not to do. This matter of the various operations is discussed in the next chapter.

of mechanical drill. The interest afforded by the objects remains, after all, external and adventitious to the numbers themselves. In the number, as number, there is no variety, but simply the ever-recurring monotony of ringing the changes on one and two and three, etc. Moreover, the appeal is constantly made simply to the memorising power. These combinations are facts to be learned. All the emphasis is laid upon the products, upon the accumulation of the information that 2 plus 2 equal 4, $2 \times 3 + 1 = 7$, $1\frac{2}{5} \times 5 = 7$, etc.; as a result, the "numbers" remain something external to the mind's own activity; something impressed upon it, and carried by it, not something growing out of its own action and coming to be a normal habit of intrinsic mental working.

Contrast with the True Method.—For the sake of indicating more clearly the defects of this method, let us follow out the contrast with the true or psychological method:

(*a*) The emphasis is all the time upon the performance of a certain mental process; the product, the particular fact or item of information to be grasped, is simply the outcome of this process. There is a given whole to be counted off into minor wholes; a group of objects to be marked off into sub-groups; a given magnitude of surface to be cut up into equal minor units of surface; a weight to be measured through equalising it with a number of sub-units of weight, etc. Then the number of sub-groups, minor unities or parts, has to be counted up in order to find the numerical value of the original whole. The entire interest is in the actual process of distinguishing the whole into its parts, and combining the parts so as to make up the value of the whole. Wherever there is a break with the mind's own activity, there the facts or principles learned are external, and interest must be partial and defective. The operation becomes mechanical, and the operator a mere machine; or else it is maintained only by a series of artificial stimulations, which keep the mind in a condition of strain—an effort which has its sole source in the need of covering the gap between the intrinsic mental activity and the abnormal action which is forced upon the mind. Wherever there is intrinsic mental activity there is interest; interest is nothing but the consciousness arising from normal activity.* Besides, this activity of parting and wholing, of measuring off into minor units of value, and summing up these minor units into the one whole, can not be performed without the mind's getting the information needed, that, e. g., $1+1+1+1$, or $1+1+2$, etc., $=4$.

(b) The appeal, according to the psychological method (number as mode of measurement), is not to memory or memorising, but is a training of attention and judgment; and this training, which forms the *habit* of definite analysis and synthesis, forms the habit of the rhythmic balancing of parts against one another in a whole, and the habit of the rhythmic or orderly breaking up of a whole into its definite parts. So far as this habit is formed the

---

* Wherever we have to appeal to external stimulus to make a subject interesting, it indicates, of course, that the activity if left to itself would cease, that the mind would wander or become listless. This means that there is no intrinsic interest, no spontaneous movement, no self-developing energy in the mind. Wherever there is this intrinsic activity, the subject is interesting of necessity and does not have to be made so.

memory will take care of itself. The facts do not need to be seized and carried by sheer effort of memory, but are reproduced, whenever needed, out of the mind's own power. The learning of facts, the preservation and retention of information, is an outcome of the formation of habit, of the attainment of power. The method which neglects the measuring function of number can not possibly lead to a definite habit; it can result only in the ability to remember.

There is no question here about the need of drill, of discipline, in all instruction. But there is every question about the true nature of drill, of discipline. The sole conception of drill and of discipline which can be afforded by the rigid unit method is that of ability to hold the mind fixed upon something external, and of ability to carry facts by sheer force of memory. By the psychological method of treating the unit as means to an end, a basis of measurement, the discipline consists in the *orderly and effective direction of power already struggling for expression or utterance.* One is the drill of a slave to fit him for a task which he himself does not understand, and which he does not care for in itself. The other is the discipline of the free man in fitting him to be an efficient agent in the realization of his own aims.

(c) Finally, the fixed unit method deadens interest and mechanizes the mind in not allowing free play to its tendencies to variety, to continual new development. As already said, according to the Grube method, the fact that $2 + 2 = 4$ always remains precisely the same, no matter how much its monotony is disguised by permutations with blocks, shoe pegs, pictures of birds, etc.

According to the measuring method, the habit or general direction of action remains the same, but is constantly differentiated through application to new facts. According to the Grube method unity is one *thing*, and that is the end of it. According to the measuring method unity may be 12 (the dozen oranges as measured by the particular orange, the day as measured by the hour, the foot as measured by the inch, the year as measured by the month, etc.), or it may be 100—e. g., the dollar as measured by the cent.

Instead of relying upon a minute and exhaustive drill in numbers from 1 to 5, allowing next to no spontaneity, severing nearly all connection with the child's actual experience, ruling out all variety as diametrically opposed to its method, it can lay hold of and give free play to any and every interest in a whole which comes up in the child's life. Unity as 12, as a dozen, is likely to be indefinitely more familiar and interesting to a child than 7; the desire to be able to tell *time* comes to be an internal demand, etc. But the Grube method must rule out 12. Twenty-five as a unity (of money, the quarter-dollar), 50 (as the half-dollar), 100 (as the dollar), are continual and lively interests in the child's own activities. Each of these is just as much one as is one eye or one block, and is arithmetically a very much better type of unit than the block by itself, because it is capable of definite measurement or rhythmic analysis into sub-units, thus involving division, multiplication, fractions, etc.—operations which are entirely external and irrelevant to the fixed unit.

Some will probably say, "But 100, or even 12, is altogether too complex and difficult a number for a child to grasp." Yes, if it is treated simply as an accumulation or aggregation of individual separate fixed units. Very few adults can definitely grasp 100 in that sense. The Grube method, proceeding on the basis of the separate individual thing as unit, is quite logical in insisting upon exhausting all the combinations of all the lower numbers. No, if 100 is treated as a natural whole of value, needing to be definitely valued by being measured out into sub-units of value. One dollar is *one*, we repeat, as much as one block or one pebble, but it is also (which the block and pebble as fixed things are not) two 50's, four 25's, ten 10's, and so on.

It may be well to remind the reader that while we are dealing here only with the theory of the matter, yet the successful dealing with such magnitude as the dozen and the dollar is not a matter of theory alone. Actual results in the schoolroom more than justify all that is here said on grounds of psychology. During the six months in which a child is kept monotonously drilling upon 1 to 5 in their various combinations, he may, as proved by experience, become expert in the combination of higher numbers, as, for example, 1 ten to 5 tens, 1 hundred to 5 hundred, etc. If the action of the mind is judiciously aided by use of objects in the measuring process which gives rise to number, he knows that 4 tens and 2 tens are 6 tens, 4 hundred and 2 hundred are 6 hundred, etc., just as surely as he knows that 4 cents and 2 cents are 6 cents; because he knows that 4 units of measurement of any kind and 2 units of the same kind are 6 units of the same kind. Moreover, this introduction of larger quantities and larger units of measurement saves the child from the chilling effects of monotony, maintains and even increases his interest in numerical operations through variety and novelty, and through constant appeals to his actual experience.

Many a child who has never seen "four birds sitting on a tree and two more birds come to join them, making in all six birds sitting on the tree," has heard of one of his father's cows being sold for $40 (4 tens), and another for $20 (2 tens), making in all 6 tens or $60; or of one team of horses being sold for 4 hundred dollars, and another for 2 hundred dollars, in all 6 hundred dollars.

When it is urged that these higher numbers are beyond the child's grasp, what is really meant? If the meaning is that the child can not picture the hundred, cannot visualise it, this is perfectly true; but it is about equally true in the case of the adult. No one can have a perfect mental picture of a hundred units of quantity of any kind. Yet we all have a conception of a hundred such units, and can work with this conception to perfectly certain and valid results. So a child, getting from the rational use of concrete objects as symbols of measuring units the fact that 4 such units and 2 such units are 6 such units, gets a clear enough working conception for any units whatever. The opposite assumption proceeds from the fallacy of the fixed unit method, and from the kindred fallacy that to know a quantity numerically we must mentally image its numerical value; grasp in one act of attention all the measuring parts contained in the quantity. Neither adult nor child, we repeat, can do this. We can not visualise a figure of a thousand sides—perhaps but few of us can " picture " one of even ten sides—but we nevertheless know the figure, have a definite conception of it, and with certain given conditions can determine accurately the properties of the figure. The objection, in short, proceeds from the fallacy that we know only what we see; that only what is presented to the senses or to the sensuous imagination is known; and that the ideal and universal, the product of the mind's own working upon the materials of sense perception, is not knowledge in any true sense of the word.

To sum up: One method cramps the mind, shutting out spontaneity, variety, and growth, and holding the mind down to the repetition of a few facts. The other expands the mind, demanding the repetition of *activities*, and taking advantage of dawning interest in every kind of value. One method relies upon sheer memorising, making the "memory" a mere fact-carrier; the other relies upon the formation of habits of action or definite mental powers, and secures memory of facts as a product of spontaneous activity. One method either awakens *no* interest and therefore stimulates no developing activity; or else appeals to such extrinsic interest as the skilful teacher may be able to induce by continual change of stimulus, leading to a varying activity that produces no unified result either in organising power or in retained knowledge. The other method, in relying on the mind's own activity of parting and wholing—its natural functions—secures a continual support and re-enforcement from an internal interest which is at once the condition and the product of the mind's vigorous action.

CHAPTER VI.

THE DEVELOPMENT OF NUMBER; OR, THE ARITHMETICAL OPERATIONS.

Numerical Operations as External and as Intrinsic to Number.

ADDITION, SUBTRACTION, MULTIPLICATION.—As we have already seen, number in the strict sense is the measure of quantity. It definitely measures a given quantity by denoting how many units of measurement make up the quantity. All numerical operations, therefore, are phases of this process of measurement; these operations are bound together by the idea of measurement, and they differ from one another in the extent and accuracy with which they carry out the measuring idea.

As ordinarily treated, the fundamental operations—addition, subtraction, etc.—are arithmetically connected but psychologically separated. Addition seems to be one operation which we perform with numbers, subtraction another, and so on. This follows from a misconception of the nature of number as a psychical process. Wherever one is regarded as one *thing*, two as two *things*, three as three *things*, and so on, this thought of numerical operations as something externally performed upon or done with existing or ready-made numbers is inevitable. Number is a fixed external something upon which we can operate in various ways: it simply happens that these various ways are addition, subtraction, etc.; they are not intrinsic in the idea of number itself.

But if number is the mode of measuring magnitude —transforming a vague idea of quantity into a definite one—all these operations are internal and intrinsic developments of number; they are the growth, in accuracy and definiteness, of its measuring power. Our present purpose, then, is to show how these operations represent the development of number as the mode of measurement, and to point out the educational bearing of this fact.

The Stages of Measurement.—We have already seen that there are three stages of measurement, differing from one another in accuracy and definiteness. We may measure a quantity (1) by means of a unit which is not itself measured, (2) with a unit which is itself measured in terms of a unit homogeneous with the quantity to be measured, (3) with a unit which is not only defined as in (2), but has also a definite relation to some quantity of a different kind. If, for example, we count out the number of apples in a peck measure, we are using the first type of measurement; there is no minor unit homogeneous with the peck measure by which to define the apple. If we measure the number of pounds of apples, we are using the second type of measurement; each apple may itself be measured and defined as so many ounces, and as therefore capable of exact comparison with the total number of pounds. In this case we have a continuous scale of homogeneous measuring units—drachms, ounces, etc.; in the first type of measuring we have not such a scale.

Finally, the pound itself may be defined not only as 16 ounces, but also as bearing a relation to some other standard; as, e. g., a cubic foot of distilled water at the temperature of 39·83 weighs 62½ pounds, the linear foot itself being defined as a definite part of a pendulum which, under given conditions, vibrates seconds in a given latitude (see page 46).

The Specific Numerical Operations.—The fundamental operations, as already said, are phases in the development of the measuring process.

1. We have seen that the comparison of two quantities in order to select the one fittest for a given end not only gives rise to quantitative ideas, but also tends to make them more clear and definite. Each of the quantities is at first a vague whole; but one is longer or shorter, heavier or lighter, in a word, *more* or *less* than the other. Here we have the germinal idea of addition and subtraction. The difference between the quantities will be a vague muchness, just as the quantities themselves are vague, and will become better defined just as these become better defined. This better definition arises with the first stage of measurement—that of the undefined unit. We begin with measuring a collection of objects by counting them off, and this suggests the measuring of a continuous quantity in a similar way—that is, by counting it off in so many paces, hand-breadths, etc. Now, in the use of the inexact unit there is given a more definite idea of the quantities and of the *more* or *less* which distinguishes them, but no explicit thought of the ratio of one to the other;

there is a counting of *like* things but not of *equal* things. In other words, the process of counting with an unmeasured unit gives us arithmetically *Addition* and *Subtraction*. The result is definite simply as to *more* or *less* of magnitude. It shows how many more coins there are in one heap than in another, how many more paces in one distance than another, in, etc. It gives an idea of the *relative value* in *this one point of moreness* or *aggregation*, but it does not bring into consciousness what multiple, or part, one of the quantities is of the other, or of their difference. This is a more complex conception, and so a later mental product.

2. With the development of the idea of quantity in fulness and accuracy the second stage of measurement is reached, in which the measuring unit is uniform and defined in terms homogeneous with the measured quantity.

This principle of measuring with an exact unit— i. e., a unit which is itself made up of minor units in the *same* scale—gives rise to *Multiplication* and *Division*, and is in reality the principle of ratio. In the addition or subtraction of two quantities we are not conscious of their ratio; we do not even *use* the idea of their ratio. In multiplication and division we are constantly dealing with ratio. We do not discover merely that one quantity is more or less than another, but that one is a certain part or multiple of another. When, for example, we multiply $4 by 5 we are using ratio; we have a sum of money measured by 5 units of $4 each, where the number 5 is the ratio of the quantity measured to the measuring unit. In division, the inverse of multiplication, ratio is still more prominent.

THE ARITHMETICAL OPERATIONS. 97

The idea of ratio involved in multiplication and division is a much more practical one than that of mere aggregation (more or less) involved in addition and subtraction, because it helps to a more accurate adjustment of means to end. Suppose a man in receipt of a certain salary knew that the rent of one of two houses is $100 a year more than that of the other, but could not tell the ratio of the $100 to his salary, it is obvious that he would have but little to guide him to a decision. But if he knows that $100 is one fifth or one fiftieth of his entire income, he has clear and positive knowledge for his guidance.

With ratio—multiplication and division—go the simpler forms and processes of fractions.*

3. The principle of measuring one scale in terms of another gives us arithmetically *proportion*, and the operations involving it, such as percentage and multiplication and division of fractions, and brings out the idea of the equation.

THE ORDER OF ARITHMETICAL INSTRUCTION.—We have already seen one fundamental objection to the ordinary method of teaching number, whether as carried on in a haphazard way or by what is known as "the Grube" method; it takes number to be a fixed

---

* In external form, but not in internal meaning, other fractions belong here also. For example, the ratio of 14 to 3 may be written $14 \div 3$ or $\frac{14}{3}$: in any case, the idea is to discover how many units of the value of 3 measure the value of 14 units, but the very fact that 3 is taken as the unit shows the *meaning* to be the discovery of the ratio of 14 to 3 as *unity*. Whether the *result* can actually be written in integral form or not is of no consequence in principle, so long as the process is the attempt to discover the ratio to unity; the process is $\frac{1}{3}$ of 14.

quantity, instead of a *mental operation* concerned in measuring quantity. We can now appreciate another fundamental objection: it attempts to teach all the operations *simultaneously*, and thus neglects the fact of growth in psychological complexity corresponding to the development of the stages of measurement. It takes each number as an entity in itself, and exhausts all the operations (except formal proportion *) that can be performed within the range of that number. It assumes that the logical order is the order of growth in psychological difficulty. *All operations are implied even in counting, but are not therefore identical.*

Logically, or as processes, all operations are *implied*, even in counting. To count up a total of four apples involves multiplication and division, and thus ratio and fractions. When we have counted 3 of the 4 apples, we have taken a first 1, a second 1, and a third 1— that is, a total of three 1's—out of the 4 which compose the original quantity. We have *divided* the original quantity of apples into partial values as units, and have taken one of those units so many times; this is multiplication. But it does not follow that, because the operations are logically implied in this process, they are therefore the same in their complete development and all equal in point of psychological difficulty; much less that they should be definitely evolved in consciousness and all taught together. The acorn implies the oak, but the oak is not the acorn. Multiplication is im-

---

* Why not proportion, or even logarithms, on the principle that everything that is logically correlative should be taught at once? The logarithm is just as much involved in say 8, as are all the multiplications and additions which can be deduced from it.

plied in the simple act of counting, and has its genesis in addition; but multiplication is not merely counting, nor is it identical with addition. The operation indicated in $\$2 + \$2 + \$2 + \$2 = \$8$ may be performed, and in the initial stages of mental growth is performed, without the conscious recognition that eight is four times two. The latter is implied in the former, and in due time is evolved from it; but for this very reason it is a later and more complex conception, and therefore makes a severer demand upon conscious attention. The summing process is made comparatively easy through the use of objects; it is little more than the perception of *related* things. The multiplication process is more complex, because it demands the actual use and more or less conscious grasp of ratio, or times, the abstract element of all numbers; it is the conception of the *relation* of things. We might go on adding twos, or threes, or fours, instantly merging each successive addend in the growing aggregate, and, *never returning to the addends*, correctly obtain the respective *sums* without the more abstract conception of times ever arising—that is, without ever being conscious that the "sum" is a *product* of which the times of repetition of *the addend* is one of the *factors*. Certainly this more abstract notion does not arise at first in the development of numerical ideas in either the child or the race.

If anyone still maintains that addition (of equal addends) and multiplication are identical processes, let him prove by mere summing (or counting) that the square root of two, multiplied by the square root of three, is equal to the square root of six; or find by logarithms the sum of a given number of equal addends.

Simultaneous Method not Psychological.—It seems clear, therefore, that the fundamental operations as formal processes should not be all taught together; on the other hand, rational *use* should be made of their logical and psychological correlation. It is one thing to perform arithmetical operations in such a way as to *involve* the *use* of correlative operations, and it is another thing to force these operations into consciousness, or to make them the express object of attention. The natural psychological law in all cases is first the use of the process in a rational way, and then, after it has become familiar, abstract recognition of it.

The method usually followed violates both sides of the true psychological principle. Because it treats number as so many independent things or unities, it can not mentally or by interpretation bring out how the operations are correlative with one another. It is only when the unit is treated not as one thing, but as a standard of measuring numerical values, that addition and multiplication, division and fractions, are rationally correlative. And it is because this correlation is not brought out and rationally used that—in spite of the teaching of all the operations contemporaneously—division is still a mystery and fractions a dark enigma.

Then, the common method errs in the opposite extreme by attempting to force the recognition of ratio, and fractions, into consciousness before the mind is sufficiently mature, or sufficiently exercised in the *use* of ratio, to grasp its meaning. The result of this unnatural method is that mechanical drill and memorizing, with the sure effect of waning interest and feeble thought, is forced upon the pupil. To master all the numerical operations contained in 6, 7, 8, and 9 is a slow and tedious process, and so the method is compelled in self-consistency to limit the range of numbers which are to be mastered in a given time. In reality it is easy for the mind to grasp the fact that $1 is a hundred ones, or fifty twos, or ten tens, or five twenties, long before it has exhausted all possible operations with such numbers as 7 or 11 or 18. It might, indeed, be maintained that a return to the old-fashioned ways of our boyhood, by which we soon became expert in the mechanical processes of addition and subtraction, would be preferable to this monotonous drill on "all that can be done with the numbers" from 1 to 10 and from 10 to 20 in the second year; for this new method is just about as mechanical as the old, and, while leaving the child little if any better prepared for the "analysis" of the higher numbers, leaves him also without the expertness in the operations which is essential to progress in arithmetic.

Division and Subtraction not to precede Multiplication and Addition.—On the ground that the "first procedure of the mind is always analytic," * some maintain that division and subtraction (the "analytic processes") should be taught before multiplication and addition.

But just as multiplication, definitely using the idea of ratio, is a more complex process than addition, so division, the inverse of multiplication, is more complex

---

* It might be asserted with some truth that the first procedure of the mind is synthetic: there must be a "whole"—a synthesis—however vague, for analysis to work upon. Certainly the *last* procedure of the mind is "synthetic."

than subtraction, the inverse of addition. We may, as we have seen, add a number of threes, for example—giving each addend a momentary attention, and then dropping it utterly from consciousness—without grasping the *factor*, which, with three as the other factor, will give a *product* equal to the *sum* of the addends. So in division, the inverse operation, this *factor* does not come merely from the successive subtractions of three from the sum until there is no remainder; here, as in addition, a further mental operation is necessary before the factors are discovered—that of counting the times of repetition; i. e., of finding the ratio of the sum (dividend) to the repeated subtrahend.

We are told, too, that when we separate 8 cubes into 4 equal parts it is instantly seen that 8 contains 2 four times, that 2 is one fourth of 8, that 2 may be taken four times from 8, and that these results being obtained independently of addition and multiplication, division and subtraction may be taught first.

There seems to be a fallacy lurking here. We may, indeed, separate 8 cubes into two parts, or four parts, or eight parts; but that is mere physical separation. Granting recognition of the concrete (spatial) element —the measuring units—how does the abstract element —the idea of times—arise? How do we know that there is four *times* 2 or eight *times* 1? Only by counting, by relating, by an act of synthesis—the last procedure of the mind in a complete process of thought. Thus, the fallacy referred to ignores one of the two necessary factors (relation) in the psychical process of number. It must presuppose that counting does not imply addition and multiplication. What is counting but addition by ones? What is five, if not one more than four; and four, if not one less than five? How is four, e. g., defined except as that number which, applied to a unit of measure, denotes a quantity consisting of three such units and one unit more? This counting, which begins with discrete quantity (collection of objects) in the first stage of measurement, is addition (with subtraction implied) by ones, and the idea of multiplication and division involved in it becomes *evolved* (in counting with an exact unit of measure) with the growth of numerical abstraction and the consequent development of the measuring power of number.

It seems plain, then, that in the development of number as the instrument of measurement there is first the rational use, leading to conscious recognition, of the aggregation idea—that is, addition and subtraction; then the definite use, leading to conscious recognition, of the factor (times) idea—that is, multiplication and division. In other words, the psychological order as determined by the demand on conscious attention is the old-time arrangement—Addition and Subtraction, Multiplication and Division.

It is the order in which numerical ideas and processes appear in the evolution of number as the instrument of measurement; the order in which they appear in the reflective consciousness of the child; the order of increasing growth in psychological complexity. This order may be said to reverse the order of logical dependence, but the psychological order rather than that of logical dependence is to be the guide in teaching.

Not Exclusive Attention to One Rule or Process.— But the true method, as based on this psychological order of instruction, by no means implies that addition and subtraction are to be completely *mastered* before the introduction of any multiplication or division or fractions. Quite the contrary. On account of their greater complexity the higher processes are not to be taught analytically—made, that is, an object of conscious attention from the first; but they may and should be freely *used*, and thus relieve the monotony of too much addition and subtraction, and at the same time prepare the way for their conscious (analytic) use.

Because of the rhythmic character of multiplication such forms of it as can be objectively presented in simple constructions—the putting together of triangles, squares, cubes, etc., to make larger or more complex figures, of dimes to make dollars—are much more easily learned than many of the addition and subtraction combinations. The ideas of ratio should be incidentally introduced in connection with certain values (e. g., 9, 12, 6, 16, 100, etc.) practically from the beginning; and consequently the process of fractions in simple forms, and its symbolic statement. Nothing but the demands of a preconceived theory could so nullify ordinary common sense as to suppose that there is no alternative between *either* exhausting all operations with every number before going on to the next higher, *or* else mastering all additions and subtractions before going on to ratio—multiplication and division. Practical common sense and sound psychology agree in recommending first the *emphasis* on addition and subtraction, with incidental introduction of the more rhythmical and obvious forms of ratio, and gradual change of emphasis to the processes of multiplication and division. If the idea of number as a mode of measurement is followed, it will be practically impossible to work in any other way. Even while working explicitly with addition and subtraction—inches, feet, ounces, pounds, dollars, cents, etc.—the process of ratio is constantly being introduced. The child can not help *feeling* that 1 inch is one third of 3 inches, 10 cents (1 dime) one tenth of a dollar, etc.; and this natural growth towards the definite conception of ratio is only checked, not forwarded, by compelling a premature conscious recognition of the nature of the process.

ADDITION AND SUBTRACTION.—The general nature of these operations as concerned with measurement through the process of aggregating minor units or parts has already been dealt with. Two or three points may, however, be considered in more detail.

1. *Work from and within a Whole.*—Here, as everywhere, the idea of a magnitude—a whole of quantity—corresponding to some one unified activity should be present from the first. Some vague quantity or whole, which is to be measured by the putting together of a number of parts, alone gives any reason for performing the operation and sets any limit to it. The process of breaking up the whole into parts and then putting together these parts into a whole, measures or defines what was originally a vague magnitude and gives it precise numerical value. In dealing, say, with 6, we may begin with a figure like this ⊠.* This is a unity or whole—it is one. But its value is indefinite. The

---

* This may, of course, be constructed out of splints, or whatever is convenient.

counting off of the various sticks changes the vague unity into a measured unity, but these parts *always fall within the original unity.* There is always a sense of the whole connecting them together. If the square has already been mastered, the figure will be recognised as one 4 + one 2. Or, if one of the diagonals is changed thus ▱, it will be recognised as two triangles—that is, as one 3 + one 3. Or, of course, it may be taken all to pieces and put together again and recognised as 6 parts of the value of 1 each. Or, the pupil may be told to make "pickets" or "tents" of the figure, and, arranging them as follows, ∧ ∧ ∧, see that there are three groups of the value of 2 each.

The principle kept in mind in this instance is that of the equation and its rhythmic construction. (*a*) According to the prevalent method, six, when reached, would be simply six ones, six separate unities, that is—not, as in the foregoing illustration, six parts of unit value each. No matter how much the teacher is urged to have the pupils recognise six at a glance, and not count up the various unities in it separately, still the fact remains that it can not, by that method, be grasped as a whole; while by the psychological method it can not be grasped in any other way. (*b*) It is also, upon the psychological method, regarded as having a value equal to (measured by) its constituent minor wholes. We are always working *within* a value, simply making it more clear and definite, not blindly or vaguely from fixed unities to their accidental sum—accidental, that is, so far as the action of mind is concerned. As a result, the psychological method appeals directly to the power of breaking up a larger whole into minor wholes, and putting these together to make a larger whole. It appeals to the constructive rhythmic interest, never to mere memorizing. It gives the maximum opportunity for the exercise of power; it leaves the minimum for mere mechanical drill. Because, dealing with wholes, intuition may be used; the rationality of the principle—*the construction of a complete whole by means of partial wholes*—may be objectively seen and clearly appreciated.

It may be laid down, then, in the most emphatic terms, that the value of any device for teaching addition depends upon whether or not it *begins with a whole which may be intuitively presented*, and whether or not it proceeds by the rhythmic partition of this original whole into minor wholes, and their recombination.

2. *Use of Subtraction as Inverse Operation.*—Upon this basis the process of subtraction is always *used* simultaneously with addition. In beginning with a fixed unity, or an aggregate of such unities, the "method" may tell us to teach addition and subtraction together (or, what is really meant, one immediately after the other), but they *can not* be employed at the same time. If 1 is one thing, 2 two things, 3 three things, and so on, it requires one mental act to unite two or more such things, and notice the resulting sum; and another act to remove one or more, and note the resulting difference. But in beginning with ⊠ and noting that it is made up of ☐, or 4, and ✕, or 2, the synthesis (recognition of the whole of parts) and analysis (recognising the parts in the whole) are absolutely simultaneous. It is one and the same act ($6 = 4 + 2$), which becomes in outward statement addition or subtraction, according as the emphasis is directed upon both of the parts equally, or upon the whole and

one of the parts. If, for example, in the above instance the ☐ and the ✕ are both equally familiar, then the construction would probably appeal to the child as addition, putting together the more familiar to make the more unfamiliar. But if the ☐ alone is very familiar, he might rather notice that the *difference* between the square and the original whole, namely, ✕, or 2 units.

3. *The Conscious Process of Subtraction slightly more Complex.*—The *conscious* recognition of subtraction, however, is a slightly more complex process—makes more demand upon attention—than the conscious interpretation of addition. In addition, the whole emphasis is upon the result; it is not necessary to keep the parts separate at all. The sum of 5 and 4, e. g., is first of all supplied by intuition, and where the association is complete the mind merely touches, as it were, the symbols, and the sum appears in consciousness. If, for example, we know that James and John and Peter have a certain amount of money—the undefined whole—of which James has 6 and John 8 and Peter 12 cents, we instantly merge or absorb each preceding quantity in the next greater—6, 14, 26. As soon as the two parts are added they are dropped as separate parts, the resulting whole is alone kept in mind. But in subtraction it is necessary to note both the whole and the given part, and the relation between them. If we say that of the total amount* James has 6 cents, John

---

* While it is not necessary always to introduce the idea of the total first in words, it should be done even verbally until we are sure that the child's mind always supplies the idea of a whole from and within which he is constantly working.

2 more than James, and Peter 4 more than John, then the addition problem requires the same attention to the two terms separately and to the result as is required in subtraction. There is the idea of definiteness or *relative* moreness, and not merely the idea of an aggregate moreness. Here, as in subtraction, we are approaching nearer to ratio.

Multiplication: Genesis of The Factor Idea.

We have seen that, though multiplication is not identical with addition (even with the special case of addition where the addends are all equal), it has its genesis in addition, taking its rise in *counting*, which is the fundamental numerical operation. Counting is the relating process in the mental activity which transforms an indefinite whole of quantity into a definite whole. It begins with discrete quantity, and is first of all largely mechanical—an operation with things. The child in his first countings does not consciously *relate* the things; his act is not one of rational counting. He is apt to think that the number-names are the names of things; that *three*, e. g., is not the *third* of three related things, but the *name* of the third thing; and on being asked to take up three he will fix upon the single thing which in counting was called three.

But starting with groups of objects and repeating the operations of parting and wholing, he soon begins to feel that the objects are related to one another and to the whole. This is a growth towards the true idea of number, but the idea is not yet developed. There is a relating, but not the relating which constitutes number. In the process of counting one, two, etc., getting as far as five, e. g., he is conscious that five is connected with what goes before. This perception is one of moreness or lessness, of aggregation; five is more than four, and at last, definitely, it is *one* more than four. With the continuance of the physical acts there is further growth towards the higher conception. He separates a whole into parts and remakes the whole; he combines (using intuitions) unequal groups of measuring units (e. g., 3 feet and 4 feet) to express them as *ones;* he counts by *ones, groups* of two things, of three things, etc., and at last the idea of *times*, of pure number, is definitely grasped. The "five" is no longer *merely* one more than four, it is five *times one*, whatever that *one* may be. In other words, he has passed from the lower idea to the higher; from the idea of mere aggregation to that of times of repetition; from addition to multiplication.

It is plain that there must be time for the development of this abstracting and generalizing power. In fact, the complete development of the "times" idea, this factor relation, corresponds with the stages of the measuring power of number. The higher power of numerical abstraction *is* the higher power of the tool of measurement. This normal growth in the power of abstracting and relating can not be *forced* by any—the most minute and ingenious—analyses on the part of the teacher. The learner may indeed be drilled in such analyses, and may glibly repeat as well as "reason out" the processes; just as he can be drilled to the repetition of the words of an unknown tongue, or any other product of mere sensuous association. But it does not follow that he knows number, that he has grasped the idea of *times*. The difficulty is not in the *word* times, as some appear to think; it is in the idea itself, and would not disappear even if the word were (as some propose) exorcised from our arithmetics. It has not yet been proposed to eliminate the *idea* itself—i. e., the idea of number—from the science of number.

SUMMARY.—(1) Counting is fundamental in the development of numerical ideas; as an act or operation with objects it is at first largely a mechanical process, but with the increase of the child's power of abstraction it gradually becomes a *rational* process. (2) From this (partly) physical or mechanical stage there is evolved the relation of more or less, and addition and subtraction arise—that is, e. g., five is one more than four. (3) The addition, through intuitions, of unequal (measured) quantities, which are thus conceived and expressed as a defined unity of so many ones, is an aid to the development of the times idea. (4) Continuance of such operations—appealing to both eye and ear—brings out this idea more definitely—e. g., five is not now simply one more than four, it is five times one. (5) Counting (by ones) groups of twos, threes, etc., brings out still more clearly the idea of times. (6) Through repeated intuitions, sums (the results of uniting equal addends) become *associated* with *times*, the factor idea (times of repetition) displaces the part idea (aggregation), and multiplication as distinct from addition arises explicitly in consciousness.

The Process of Multiplication.—The expression of measured quantity has, we have seen, two components, one denoting the unit of measure, and the other denoting the number of these units constituting the quantity. But since the unit of measure is itself composed of a definite number of parts—is definitely measured by some other unit—it is clear that we actually conceive of the quantity as made up of so many given units (direct units of measure), each measured by so many minor units. For convenience we may call these minor units "primary," as making up the direct unit of measure, and this direct unit, as being made up of primary units, may be called the "derived" unit. We shall thus have in the complete expression of any measured quantity, (1) the derived unit of measure, (2) the number of such units, and (3) the number of primary units in the derived unit of measure.

For example, take the following expressions of quantity: In a certain sum of money there are *seven* counts of *five* dollars each; here the derived unit of measure is *five* dollars, the *number* of them is *seven*, and the primary unit is one dollar. The cost of a farm of sixty acres at fifty dollars an acre is sixty fifties; here the derived unit of measure is fifty dollars, the number of them sixty, and the primary unit one dollar. The length of a field is fifteen chains—that is (in yards), fifteen twenty-twos; here the derived unit is twenty-two yards, the number of them fifteen, and the primary unit one yard. In the quantity $\$\tfrac{3}{4} \times \tfrac{3}{5}$ the primary unit is $\$1$, the derived unit $\$\tfrac{3}{4}$, and $\tfrac{3}{5}$ is the *number* expressing the quantity in terms of the derived unit.

Now, when a quantity is expressed in terms of the derived unit, it is often necessary or convenient to express it in terms of a primary unit. Thus, in the foregoing examples, the sum of money expressed as seven fives may be expressed as *thirty-five* ones; the cost of the farm, expressed as sixty fifties, may be expressed as *three thousand* ones; and the length of the field, expressed as *fifteen* twenty-twos, may be expressed as *three hundred and thirty* ones (yards); and $\frac{22}{4} \times \frac{3}{5}$ is measured by $\frac{9}{20}$ in terms of the primary unit. In each of these cases the second expression of the measured quantity merely states explicitly the number of minor (or primary) units which is *implied* in the first expression. The operation by which we find the number of primary units in a quantity expressed by a given number of derived units is *Multiplication*. It is plain that the idea of *times* (pure number, ratio) is prominent in this operation; we have the *times* the primary unit is taken to make up the derived unit, and the *times* the derived unit is taken to make up the quantity. The multiplicand always represents a number of (primary) units of quantity; the multiplier is always pure number, representing simply the times of repetition of the derived unit. But from the nature of the measuring process the two factors of the product may be interchanged, the times of repetition of the primary unit may be commuted with the times of repetition of the derived unit; in other words, the *number* which is applied to the primary unit may be commuted with the *number* which is applied to the derived unit.

Correlation of Factors.—In our conception of measured quantity these two ideas are, as has been shown, absolutely correlative. Measuring a line of *twelve* units by a line of *two* units, the numerical value is *six*; if we consciously attend to the process, the related conception instantly arises; we can not think six times two units without thinking two times six units, because we can not think one unit six times without thinking *one* whole of six units. So, in measuring a rectangle 8 inches long by 10 inches wide, we can not analytically attend to the process which gives the result of 8 square inches taken ten times without being conscious of the inevitable correlate, 10 square inches taken eight times. In general: To *think* the measurement of any quantity as b units taken a times, is to think its correlate a units taken b times; for b units is b times *one* unit, and every one in b is repeated a times, giving a units once, a units twice, etc.—that is, a units b times.

Educational Applications.

1. Just as, in addition, we must always begin with a vague sense of some aggregate, and then go on to make that definite by putting together the constituent units, while in subtraction we begin with a defined aggregate and a given part of it, and go on to determine the other parts; so, in multiplication, we begin with a comparatively vague sense of some whole which is to be more exactly determined by the "product," while in division we begin with an exactly measured whole, and go on to determine exactly its measuring parts. In multiplication the order is as follows: (1) The vague or imperfectly defined magnitude; (2) the definite unit of value (primary unit), which has to be repeated to make the derived (direct) unit of measure—the multiplicand; (3) the number of times this derived unit is to be repeated —the multiplier; and (4) the product—the vague magnitude now definitely measured.

The operation of multiplication, therefore, already

implies division; the definite unit of measurement which constitutes the multiplicand is always a certain exact (equal) portion of some whole. Hence multiplication always implies ratio; the whole magnitude bears to the unit of measure a ratio which is expressed in the number of times (represented by the multiplier) the unit has to be taken to measure that magnitude—to give it accurate numerical value. In fact, the process is simply one of changing the *number* which measures a magnitude by changing the unit of measure—i. e., by substituting for the given unit of measure the primary unit from which it was derived.*

2. In multiplication, then, as in addition, we are not performing a purposeless operation, or one with unrelated parts and isolated units; rather, we begin and end with some magnitude requiring measurement, keeping in mind that what distinguishes multiplication is the kind of measurement it uses—that, namely, in which a unit itself measured off by other units is taken a certain number of times.

3. The psychology of number, therefore, imperatively demands that the quantity which is to be finally expressed by the "product" should first be *suggested*, just as in addition the quantity given by "sum," within which and towards which we are working, is kept in

---

* In such instances as multiply 7 apples by 4, the idea of exact division or ratio is not so evident, but the 7 apples must be taken as one of four equal portions—i. e., as having the ratio $\frac{1}{4}$ to the whole quantity. The fact, however, that the idea of an exact unit of measurement is not so clearly present, is a strong reason for using fewer examples of this sort, and more of those involving standard units of measure.

mind from the first. If the child sees, e. g., that there is a certain field of given dimensions whose area is to be ascertained, or a piece of cloth of given length and price per yard, of which the cost is to be determined, the mind has something to rest upon, a clearly defined purpose to accomplish. Beginning with a more or less definite image of the thing to be reached, the subsequent steps have a meaning, and the entire process is rational and consequently interesting. But when he is asked how much is 4 times 8 feet, or 9 times 32 cents, there is no intrinsic reason for performing the operation; psychologically it is senseless, because there is no motive, no demand for its performance. The sole interest which attaches to it is external, as arising from the mere manipulation of figures. Under an interested teacher, indeed, even the pure " figuring " work may be interesting; but this interest is re-enforced, transformed, when the mechanical work is felt to be the means by which the mind spontaneously moves by definite steps towards a definite end. This does not mean, we may once more remark, that examples like 8 feet × 4, or even 8 × 4, are to be excluded, but only that the *habit* of regarding number as measuring quantity should be permanently formed. The pupil should be so trained that all addends, sums, minuends, products, multiplicands, dividends, quotients, could be instantly interpreted in their nature and function as connected with the process of measurement. For example: A farmer has 8 bushels of potatoes to sell, and the market price is 55 cents a bushel: how much can he get for them?

This and similar examples are often presented in such a way that when the pupil gets the product, $4.40, his mind stops short with the mere idea of the product as a series of figures. This is irrational; $4.40 in itself is not a product; no quantity or value is ever in itself a product; but *as* a product it measures more definitely the value of some quantity. In other words, the product must always be *interpreted;* it must be recognised as the accomplished measurement of a *measured* quantity in terms of more familiar or convenient units of measure.

4. The multiplicand must always be seen to be a *unit* in itself, no matter how large it is as expressed in minor units. It signifies the known value of the unit with which one sets out to measure; it is the measuring rod, as it were, which is none the less (rather the more) a *unit* because it is defined by a scale of parts. A foot is none the less *one* because it may be written as 12 inches or as 192 sixteenths; nor is a mile any the less a unit because it is written as 320 rods or as 5280 feet. The ineradicable defect of the Grube method, or any method which conceives of a unit as one thing instead of as a standard of measuring, is that it can never give the idea of a multiplicand as just one unit—a part used to measure a whole.

5. It is important so to teach from the beginning that a clear and definite conception of the relation between parts and times may be developed. Of course, nothing is said till the time is ripe about the law of "commutation"; but the idea should be present, and should be freely used. If a quantity of 12 units is measured by 3 units repeated *four* times, the child can be led to see—will probably discover for himself—that this measurement is identical with the measurement

4 units repeated *three* times. Rationally using this idea of commutation in repeated operations, the child will soon get possession of a principle by which he can easily interpret both processes and results in numerical work.

CHAPTER VII.

NUMERICAL OPERATIONS AS EXTERNAL AND AS INTRINSIC TO NUMBER.

Division and Fractions.

Division.—As multiplication has its genesis in addition, but is not identical with it, so division has its genesis in subtraction, but is not identical with it. Just as multiplication comes from the explicit association of the *number* of *equal addends* with their *sum*, and the substitution of the *factor* idea (ratio) for the *part* idea, so division comes, in the last analysis, from the explicit association of the number of equal subtrahends from the same sum (dividend), and the substitution of the factor idea for the part idea. In other words, division is the inverse of multiplication, just as subtraction is the inverse of addition. Further, as in multiplication, both factors are the expression of a measured quantity and are interchangeable, so in division either of the factors (divisor and quotient) which produced the dividend can be commuted with the other. In multiplication, for example, we have 4 feet × 5 = 5 feet × 4 = 20 feet; and the inverse problem in division is, given the 20 feet, and either of the factors, to find the other factor. We solve the problem not by subtraction, but by the use of the factor, or ratio, idea.

In multiplication, as already suggested, we may look at a product of two factors in two ways: For example, 20 feet = 4 feet × 5 = also 5 feet × 4, or five times four times 1 foot = four times five times 1 foot—that is, we may use the primary unit of measure "1 foot" with either the four times or the five times. Or, stated in general terms, b times a times the primary unit of measure is identical with a times b times this primary unit—that is, we may interchange at pleasure the *numerical* value of the measuring *unit* (the *derived* unit as made up of primary units) with the *numerical* value of the whole quantity as made up of these derived units. This is important as interpreting the process and result in division. If we have 20 feet and the factor 4 feet given to find the other factor, we use the measurement 4 feet × 5 = 20 feet. If, on the other hand, we have the 20 feet and the *number* 5 given to find the other factor, we may use *either* measurement; we may divide directly by the number 5, or we may "concrete" the 5 (consider it as denoting 5 feet), and get the other factor 4 (times); for we know that 4 times 5 feet is identical with 5 times 4 feet, and the conditions of the question require the latter interpretation. In other words, we first of all determine what the problem demands, times or parts, then operate with the pure number symbols, and interpret the result according to the conditions of the problem.

Illustrations of Division.—Let us take a few illustrations of these inverse operations: (1) We count out fifteen oranges, by groups of five, and the *number* of groups is *three*. We count them out in five groups, and the number of oranges in each group is *three*. These

NATURE OF DIVISION AND FRACTIONS. 121

are said to be two totally different operations; for, it is alleged, in one case we are searching for the size (the numerical value) of a group—the unit of measurement; in the other for the *number* of groups. But a little reflection will show that they are not "radically different" operations; they are psychological correlates, if not identities. In counting out fifteen oranges in groups of five there is a count of *five*, then another count of five, then another count of five, and finally a counting of the number of groups. Psychologically, in counting out five there is a mental sequence of five acts (a partial synthesis); this is repeated three times, and finally the number of these sequences is counted (complete synthesis), and found to be three. In the second case, where the number (*five*) of groups is given, we begin by putting one orange in each of five places, making, as before, a "count" of *five* oranges; this operation is repeated till all are counted out: and finally we count the number in each of the five groups. That is, there is a mental sequence of five acts, which is repeated three times, and finally the *number* of such sequences is counted in counting the number of oranges in a group. It would be hardly too much to say that these two mental processes are so closely correlated as to be identical. Neither the *three* times in the one question nor the *three* oranges in the other can be found without counting out the whole quantity in groups of five oranges each (see page 75). There is hardly a difference even in the rhythm of the mental movement. This division by counting is the actual process with things: it is the way of the child and of the savage or the illiterate man; it is exactly symbolized in the "two kinds of division"

—that by a concrete divisor when we are searching for the number of the parts as actual units of measure; and that by an abstract divisor when we are searching for the size of the parts—i. e., for the number of minor units in the actual unit of measure.

With this actual process of counting out the objects the arithmetical operation exactly corresponds. Working by long division as more typical of the general arithmetical operation, we have:

I. *Division :* 15 oranges ÷ 5 oranges; i. e., 15 oranges are to be counted out in groups of 5 oranges; how many groups?

5 oranges	15 oranges	1°—1st partial multiplier			
	5 "				
	10 "	1°—2d " "		}	= 3 times.
	5 "				
	5 "	1°—3d " "			
	5 "				

II. *Partition :* 15 oranges in 5 groups; how many in each group?

5 times	15 oranges	1 orange—1st partial multiplicand	
	5 "		
	10 "	1 " —2d " "	= 3
	5 "		oranges.
	5 "	1 " —3d " "	
	5 "		

That is, once more, both problems are solved by counting out the whole quantity in groups of *five*.

(2) Solve the following problems: (*a*) Find the cost of a town lot of 36 feet frontage at $54 a foot. (*b*) At the rate of $54 a foot, a town lot was sold for $1944,

find the number of feet frontage. (c) Find the price per foot frontage when 36 feet cost $1944.

(a)

(i) $54
      36
      ────
      1620 = 30 times $54
       324 = 6   "   "
      ────
      $1944 = 36 "   "

Or, by the correlate, $36 × 54:

(ii) $36
       54
      ────
      1800 = 50 times $36
       144 = 4   "   "
      ────
      $1944 = 54 "   "

(b)

$54)$1944
    1620 = 30 times $54
     324 = 6   "   "
    ────
     324;   36 times $54

(c) *Partition.*

36)$1944
   1800 = $50   36 times
    144 = $4   36   "
   ────
    144; $54   36 times

On comparing the successive steps in (b) with those in (a) they will be seen to correspond exactly—that is, (b) is the *exact* inverse of (a). But the steps in (c) do not correspond with those in (a), the operation is not the exact inverse of (a); it is seen to be the exact inverse of the correlative (ii) of (a). This indicates the connection of the operations through the law of commutation; and shows, once more, that either of the correlated measurements (i) and (ii) may be used in the solution of (c). It should be noted, further, that (c) is a case of so-called *partition*, yet involves a series of subtractions that is a series of partial dividends (why not *partiends?*) and partial quotients.

Not Two Kinds of Division.—From the foregoing we see that just as a product of two factors may be interpreted in two ways, so there may be two *interpretations* of the result of the inverse operation, division. The factor sought may be either the numerical value of the dividing part ("derived unit") in terms of the primary unit which measures it, or the numerical value of the quantity in terms of the derived unit. But these numerical values may be interchanged at convenience, provided the results are rightly interpreted. There are *not two* kinds of division; there is one operation leading to one numerical result having two related meanings. It seems therefore unnecessary, either on psychological or practical grounds, to institute two kinds of division—viz., division (why not *quotition ?*) in the ordinary sense of the word, and "partition"—when the search is for the *numerical value* of the measuring quantity. When the search is for the *numerical value* of the measuring unit, is not the pupil likely to become perplexed by a series of parallel definitions—of divisor, dividend, quotient—for the two divisions when he finds that the operations in both cases are exactly alike? If there is confusion in using the term division in two senses, is there not more confusion in using the two terms, divisor and quotient, each with two different meanings? Without doubt, the meaning of the result should be grasped; but this can not be done by simply giving two names to exactly the same arithmetical operation. Better give one name to one operation resulting in two correlated meanings than to have two names for one and the same operation. The new name does not help the pupil either in the numerical work or in the interpretation of the result. How is the child to know whether a given problem is a case of division or of "partition"? He can not know without an intellectual operation, analysis, by which he grasps what is given and what is wanted in the problem. In other words, he must know the meaning of the problem, must know whether it is times or measuring parts he is to search for, *before he begins the operation;* to this knowledge the different names afford him no aid whatever.*

Partition, like Division, depends on Subtractions.—It is said, indeed, that in "partition" we are searching for the numerical value of one of a given number of equal parts which measure a quantity, and as a number can not be subtracted from a measured quantity, the problem can not be solved by division. To this the answer is easy: In the first place, the divisor in the arithmetical operation can be a number, and the subtractions rationally explained (see page 122). And, besides, we can by the law of commutation concrete the number, find the related factor, and properly interpret the result. But, in the second place, if the divisor can not be an abstract number, what magic is there in a strange name to bring the impossible within the easy reach of childhood? It seems, according to the *partitionists*, that 20 feet ÷ 5 feet represents a possible and intelligible operation; but that 20 feet ÷ 5 becomes possible and intelligible only by calling the implied operation a case of "partition"; it is then simply one fifth of 20 feet—that is, 4 feet. Certainly, if we know the multiplication table, we know that one fifth of 20 feet is 4 feet, but we know equally well that 5 feet is one fourth of 20 feet. These are not typical cases for the argu-

---

* Owing to the fixed unit fallacy, the theory of the "two divisions" makes an unwarranted distinction between the actually measuring part and its times of repetition. The measuring part, as well as the whole, involves both the spatial element (unit of quantity) and the abstract (time) element; it is itself a quantity that is measured by a minor unit taken a number of times.

ment; though attention to the processes even in these cases (see page 122) will show that if 20 feet ÷ 5 is impossible because "division" is a process of subtraction, so also is the process one fifth of 20 feet, because "partition" *is equally a process of subtraction* (page 122). For example, the operation indicated in $14899623 ÷ $4681 it is admitted involves subtraction—i. e., the separation of the dividend into parts, and the obtaining of partial quotients. But it is clear that 1-4681th of this dividend (partition) is obtained by exactly the *same process*—i. e., in both cases we have a first subtraction of 3000 times the divisor, a second of 100 times, a third of 80 times, and a fourth of 3 times, getting the same numerical quotient of 3183 in both operations; but 3183 is *interpreted* as pure *number* in the first case, and as measured *quantity* in the second—the so-called partition.

In fine, when it comes to pass that there can be a clear conception of a foot as measured by inches without the thought of *both* the factors, *one inch* and *twelve times*, then, but not till then, it may be rationally affirmed that the "two divisions" are radically different and totally unrelated processes.

Fractions.

The process of fractions as distinguished from that of "integers" simply makes *explicit—especially in its notation—both the fundamental processes, division and multiplication (analysis-synthesis), which are involved in all number.*

In the fundamental psychical process which constitutes number, a vague whole of quantity is made definite by dividing it into parts and counting the parts. This is essentially the process of fractions. The "fraction," therefore, involves no new idea; it helps to bring more clearly into consciousness the nature of the measuring process, and to express it in more definite form. The idea of ratio—the essence of number—is implied in simple counting; it is more definitely used in multiplication and division, and still more completely present in fractions, which use both these operations. Fractions are not to be regarded as something different from number—or as at least a different kind of number—arising from a different psychical process; they are, in fact, as just said, the more complete development of the ideas implied in all stages of measurement. So far as the psychical origin of number is concerned, it would be more correct to say that "integers" come from fractions than that fractions come from integers. Without the "breaking" into parts and the "counting" of the parts there is no definitely measured whole, and no exact numerical ideas; the definite measurement is simply (*a*) the number of the parts taken distributively (the analysis), and (*b*) the number of them taken collectively (the synthesis). The process of forming the integer, or whole, is a process of taking a part so many times to get a complete idea of the quantity to be measured; and at any given stage of this operation what is reached is both an integer and a fraction—an integer in reference to the units counted, a fraction in reference to the measured unity.

Even in the imperfect measurement of counting with an unmeasured unit, the ideas of multiplication and division (and therefore of ratio and fractions) are implied in the operation. We measure a whole of fifteen apples by threes; we count the parts—i. e., relate or *order* them to one another, and to the whole from which and within which we are working. This counting has a double reference—i. e., to the unit of measure, and to the whole which all the units make up. When, for example, we have counted two, three, . . . we have taken one unit of measure two, three, . . . times, and each count is expressed or measured by the numbers two, three, . . . —i. e., by "integers"; but also in reaching any of these counts we have—in reference to the whole—taken *one* of the five, *two* of the five, *three* of the five, etc.; that is, one fifth of the whole, two fifths, three fifths, etc.

No Measurement without Fractions.—When we pass to measurement with exact units of measure, this idea of fractions—of equal parts making up a given whole—becomes more clearly the object of attention. The conception, 3 apples out of 5 apples (three fifths of the whole) has not the same degree of clearness and exactness as that of 3 inches out of a measured whole of 5 inches. Why? Because in the former case we do not know the exact *value*, the *how much* of the measuring unit; in the latter case the unit is exactly defined in terms of other unities larger or smaller; in 3 apples the units are *alike;* in 3 inches the units are *equal.* So in measuring a length of 12 feet we may divide it into 2 parts, or 3 parts, or 4 parts, or 6 parts, or 12 parts; then we can not really *think* of the 6 parts as making the whole without thinking that 1 is one sixth; 2, two sixths; 3, three sixths; etc. In the process of inexact measurement the idea of fractions is involved; in that of exact measurement, this idea is more clearly defined in consciousness. In short, wherever there is exact measurement there is the conception of fractions, because there is the exact idea of number as the instrument of measurement. The process of fractions, as already suggested, simply makes more definite the idea of number, and the notation employed gives a more complete statement of the analysis-synthesis, by which number is constituted. The number 7, for example, denotes a possible measurement; the number $\frac{7}{10}$ states more definitely the actual process. It not only gives the absolute number of units of measure, but also points to the definition of the unit of measure itself—that is, the 7 shows the absolute number of units in the quantity, while the 10 shows a relation of the unit of measure to some other standard quantity, a primary unit of reference by which the actual measuring unit is defined. If a quantity is divided into 2 equal parts, or 3 equal parts, or 4 equal parts, or n equal parts, the 2, 3, 4, ... n shows the entire number of parts in each measurement, and corresponds with the "denominator" of the fraction which expresses the measured quantity as *unity;* and in counting up (*e*-numerating) the parts (units) we are constantly making "numerators"—e. g., 1 out of n, 2 out of n, 3 out of n, etc.; or 1-nth, 2-$n$ths, ... n-$n$ths, or $\frac{n}{n}$, which is the measured *unity*. Or, if attention is given to the measuring units—the ones—the parts are expressed by 1, 2, 3, etc., and the measured quantity itself is expressed by $\frac{n}{1}$. Again, measuring the side of a certain room, we find it to contain $1\frac{2}{3}$ yards. This is a full statement of the process of measurement; it means (1) that the primary unit of measurement (the standard of reference) is one yard, (2) that the derived unit of measurement is one third of this, and (3) that this derived unit is taken nineteen times to measure the quantity. This is seen to agree with the mental process of the exact stage of measurement in which the unit of measure is itself defined or measured (see page 94). There must be, as we have seen, (1) a standard unity of reference (the primary unit), (2) a derived unit (the unit of direct measurement), and (3) the number of these in the quantity. The fraction gives complete expression to this process: In $\$\frac{3}{4}$, for example, (1) the dollar is the unit of reference; (2) it is divided into four parts to get the derived unit—the actual unit of measure; (3) the "numerator" 3 shows how many of these units make up the given quantity, and expresses the *ratio of this quantity to the standard unity.*

So, again, the measurement—19 feet—of the side of a room can be stated in terms of other units of the scale. It is 12×19 inches, or $19 \div 3$ yards, and the first of these expressions, as well as the second, is one of fractions; it is $\frac{228}{1}$—that is, not 228 *ones* merely, but 228 of a definite *unit* of measure—namely, one twelfth of a foot; just as the second is $\frac{19}{3}$—i. e., 19 times a unit of measure defined by its relation to the yard. In the former case we do not generally state the measurement in fractional form, but the *interpretation* of it demands an *explicit reference to a denominator.* Note what this brings us to: 19 (feet) = 228 (inches) = $\frac{19}{3}$ (yards) = $\frac{209}{11}$ (rods)—that is, four entirely different numbers equal to one another; a result which must appear utterly meaningless to a child who has been trained by the fixed unit method. Any method which treats number as a name for physical objects can not but reach just such absurdities. Only the method which recognises that number is a psychical process of valuation (analysis-synthesis) is free from such difficulties. The unit does not designate a fixed thing; it designates simply the unit of valuation, the how much of anything which is taken as *one* in measuring the value (or how much) of a group or unity. It defines how many units each of so much value make up the so much of the whole. The complete process is one of fractions, and the full statement of it is a fraction, whether written out in full or necessarily understood in the interpretation. The 228 inches is $\frac{228}{1}$, signifying that the number of the derived units of measure in one inch is 1; 19 feet is $\frac{19}{3}$ yards, signifying that the number of the derived units of measurement in one yard is 3; the $\frac{209}{11}$ rods show that the number of units of measurement in one rod is 11; in other words, the unit of measure in $\frac{19}{3}$ is *one* of the three equal parts of one yard, etc.

It appears, therefore, that every numerical operation which makes a vague quantity definite, when fully stated, involves the "terms" of a fraction—that is, a fraction may be considered as a convenient language (notation) for expressing quantity in terms of the process which measures or defines it—which makes it "number."

A fraction, then, completely defines the unity of reference, and thus determines the *unit* of measure for the quantity that is to be measured. Thus the inch may be defined from $\frac{1\frac{2}{3}}{12}$ foot, the foot from $\frac{3}{3}$ yard, the ounce from $\frac{16}{18}$ pound, the cent from $\frac{100}{100}$ dollar, etc. In each case the denominator shows the analysis of a standard unity into units of measurement—i. e., the unity in terms of the units taken collectively. Thus the measurements of the quantities 7 inches, 5 ounces, 35 cents are more explicitly stated by the respective fractional forms $\frac{7}{12}$ foot, $\frac{5}{16}$ pound, $\frac{35}{100}$ dollar, because the unit of measure in each case is consciously defined by its relation to a standard unity in the same scale.

It is clear that the definition of number (page 71) includes the fraction, for in both fraction and integer the fundamental conception is that of a quantity measured by a number of defined parts—the conception of the ratio of the quantity to the measuring unit. The fraction differs from the integral number—in so far as it differs at all—in defining the measuring unit, and thus giving more completely the psychical operation in the exact stage of measurement.

If the fraction, as being a number, is a mode of measurement, there appears to be no need of a special definition of it as the foundation of a new or different class of numerical operations. The definitions which ignore fractions as a mode of measurement are in general vague and inaccurate, and lead to much perplexity in the treatment of fractions. It is hardly accurate to say that a "fraction is a number of the equal parts of a unit," or that "it originates in the division of the *unit* into equal parts." Here the important distinction between unity and a unit is overlooked. Measuring a piece of cloth we find it contains four yards: before measurement it was mere unity, after measurement a defined unity; but in neither case is it a *unit*. It is, after measurement, a unity of *units*—a sum. Nor is it entirely consistent with the measuring idea to say that a fraction is one or more of the equal parts of a unity. Of course, in counting the equal parts of a measured whole —a unity—we take a number of parts in making the synthesis of *all* the parts. But since a fraction is a number, and therefore denotes measured quantity, it denotes a whole quantity, a unity—e. g., $\frac{4}{3}$ of a yard is as much a quantity—a measured *unity*—as 4 yards or 40 yards; it is a fraction in its relation to a larger unity, the yard taken as a standard of reference.

The Improper Fraction.—From the same misapprehension of the nature of number endless discussions arise regarding the classes of fractions "proper," "improper," etc. With a right conception of the measuring function of a fraction there is no mystery about the "improper" fraction. From the definition of a fraction as a "number of the equal parts of a unit," it is inferred, e. g., that $\frac{4}{3}$ of a yard can not be a fraction, because it represents not *parts* of a unit, but the *whole* unit and something more. Since 3 thirds make up the yard (the unit), whence come the 4 thirds?

In this objection we have the fallacy of the fixed unit as well as the misapprehension of the nature of number. The fraction in the expression $\frac{4}{3}$ of a yard is a number. It means the repetition of a unit of measure to equal a certain quantity. This unit of measure is not the yard; it is a unit defined by its relation to the yard; it is one of the three equal parts into which the unity yard is divided to get the direct unit of measure; and there is absolutely nothing to make the yard the limit of quantity to which this unit can be applied. The yard is the primary unit of reference from which the actual measuring unit is derived, and there is no more mystery in the application of this unit to measure a quantity greater than the primary unit than in the measured quantity, 3 feet × 4, because it is greater than *one foot*, the primary unit from which the measuring unit (3 feet) is derived.

The expression 4 thirds of a yard indicates an exactly measured quantity; exactly measured, because the unit of measure is itself measured in its relation to another quantity of the same scale. This properly defined unit (1 third of a yard) can be applied to any homogeneous quantity whatever, and may be contained in such quantity one, two, three, four, . . . n times; in fact, 4 thirds yard, 5 thirds yard, . . . n thirds yards are only different and more exact ways of stating the measurements—4 feet, 5 feet, . . . n feet.

The Compound Fraction.—Nor is there any difficulty in interpreting a "compound" fraction. The value of 8 yards of cloth at $\$\frac{3}{4}$ a yard is expressed by $\$\frac{3}{4} \times 8$, a measurement which ought to occasion no more perplexity than $\$3 \times 8$, when it is understood that the denominator merely defines the unit of measure with reference to the primary unit. So the value of $\frac{3}{4}$ yard of cloth at $\$8$ a yard is expressed by $\$8 \times \frac{3}{4}$, a measured quantity where, once more, the denominator shows how the unit of measure is to be obtained—i. e., it shows which of the myriad ways of parting and wholing $\$8$—the unity of reference—will give the direct or absolute unit of measure. This explanation applies to $\$\frac{8}{25} \times \frac{3}{4}$, and to any compound fraction whatever.

The Complex Fraction.—It is said that the complex fraction is an impossibility, because a quantity can not be divided into a fractional number of equal parts—e. g., if the denominator of such a fraction is $\frac{3}{4}$, it implies the division of some unity into 3 fourths equal parts, which is absurd. This is to restrict the term fraction by the imperfect definition already quoted, which ignores number as measurement and fractions as an explicit statement of the measuring process. Division and multiplication are fundamental in the psychical process of defining quantity; the fraction simply brings the process articulately into consciousness, and by its notation gives it complete expression. The statement that the fraction process and the division process are totally distinct is so far from being true, that there is no division without the fraction idea, and no fraction without the division idea. Both are identified by the law of commutation—a law which is the expression of a necessary and universal action of the mind in the measurement of quantity. The symbol $\frac{3}{4}$ foot is an exact expression for a measured quantity; like every other such expression, it defines the unit of measure and denotes the times this unit is repeated; and, like every such expression, it has two interpretations corresponding to the related conceptions of the measured quantity: it is $\frac{1}{4}$ foot × 3, or 3 feet × $\frac{1}{4}$. We shall be justified in treating these two things (the fraction and division) as entirely distinct when we are able to conceive that 3 feet × 4, and 4 feet × 3 are unrelated measurements of totally different quantities.

It may be noted, then, that in the "complex" fraction just as in division there may be two interpretations. In $12 ÷ $4 the measuring is not by *four* parts—it is a parting by *fours*; while in $12 ÷ 4 there is a measuring by four parts—it is a parting by *threes*. But in every case, as has often been shown, *either* the size of the parts is given to find their number, *or* the number of the parts to find their size. The same thing holds in so-called complex fractions. In $\frac{\$9}{\$\frac{3}{4}}$—as, e. g., find how much cloth at $\$\frac{3}{4}$ a yard can be bought for $9—it is not proposed to divide $9 into $\$\frac{3}{4}$ equal parts, but to find the times the measuring unit $\$\frac{3}{4}$ is taken to make $9. Nor in $\frac{\$9}{\frac{3}{4}}$—as, e. g., find cost per yard when $9 was paid for $\frac{3}{4}$ yard—is there any attempt to divide $9 into 3 fourths equal parts. The purpose is to find the quantity which, with $\frac{3}{4}$ as multiplier (i. e., taken $\frac{3}{4}$ times), will give $9; and it is a matter of indifference whether the expression is called a "complex" fraction or division of fractions, for fractions are necessarily correlated with multiplication *and* division by the uniform action of the mind in dealing with quantity.

Summary and Applications.

1. All numerical operations are intrinsically connected with number as measurement, and distinguished from one another through the development of the measuring idea in psychological complexity. Addition and subtraction have their origin in the operation of counting with an unmeasured unit—they do not explicitly use the idea of ratio, but merely that of more or less—the idea of aggregation. Multiplication and division have their origin in the use of an exact unit of measure—a unit which is itself defined—and,

besides the idea of aggregation, use the idea of ratio. It follows, accordingly, that addition and subtraction should precede in order of formal instruction, multiplication, and division. Addition and subtraction, being inverse operations, should go together, with the emphasis at first slightly upon addition.

2. Multiplication and division, being inverse operations, should go together, with the emphasis first upon multiplication. Multiplication should not be taught as a case of addition, nor division as a case of subtraction. But the factor idea (ratio or number) should in each case displace the idea of aggregation. While this is the order of analytical instruction, the processes involved in multiplication should be used—that is, in primary teaching there should be frequent excursions into these processes in accordance with the fundamental psychological law : "First the rational use of the process, and ultimately conscious recognition of it."

3. In multiplication the multiplicand, strictly speaking, always represents a measured quantity, and is commonly said to be "concrete"; the multiplier always represents pure number—the ratio, in fact, of the product to the multiplicand. But, as the multiplicand always involves the idea of number (it expresses the *number* of primary measuring units), the two factors of the product may be interchanged—that is, the multiplier may be made the concrete quantity, and the multiplicand the pure number denoting times of repetition.

4. Division is the inverse of multiplication. We have the product given and one of the two factors which produce it to find the other factor. And since there are two interpretations of the process of multipli-

cation there may be two interpretations of its inverse process, division. But there are not two kinds of multiplication nor two kinds of division. In each case there is one process and one result with two interpretations. No assistance to this interpretation can be afforded by giving the name "partition" to the process by which we find the size, or numerical value, of the measuring part. The student knows what he is looking for before he begins the operation—whether for the *value* of the parts in terms of the primary unit, or the *number* of them in the whole quantity. It is not necessary for him to give a new name to an old operation. Besides, if he does not know what he is searching for before he begins the numerical work, the new name throws no light upon the subject.

From the relation existing between multiplication and division, it is seen that in division the dividend—or multiplicand, as being the product of two factors—always represents a measured quantity—i. e., it is concrete; the divisor may denote either a concrete quantity or a pure number; and the quotient is of course numerical in the one case and interpreted as concrete in the other.

5. In fractions there are no mental processes different from what are involved in number as a mode of measuring quantity. The psychical process by which number is formed is from first to last essentially a process of "fractioning"—making a whole into equal parts and remaking the whole from the parts. In the process of number we start with a whole; we have a unit of measurement; we repeat the unit of measurement to make up the whole. In a measured quantity represented by a fraction we do exactly the same thing. We

NATURE OF DIVISION AND FRACTIONS.

begin with a whole of quantity; we use a unit of measure of the same kind as the quantity; we repeat the unit of measure to make up the whole. The fraction by its notation brings out more explicitly the actual process of measurement—that is, it not only gives the number of units of measure, but actually defines the unit itself in terms of some other unit in the same scale. In other words, a *fraction sums up in one statement the mental process of analysis-synthesis by which a vague whole is made definite.*

1. As fraction at all it expresses a portion of some group or whole with which the quantity represented by the fraction is compared, and which defines the measuring unit. Thus, $\frac{1}{8}$ yard of cloth is itself a whole, a definitely measured quantity; but it is a fraction as regards the standard of reference, *yard*, which defines the direct measuring unit, one eighth of a yard.

2. A fraction, therefore, always denotes (*a*) the absolute number of units in a measured quantity; (*b*) the number of such units in some standard quantity which defines the measuring unit in (*a*); and (*c*) the ratio of the given quantity (represented by the fraction) to this standard of reference. The numerator of the fraction gives (*a*) and (*c*), and the denominator gives (*b*). Of course, any part (or multiple) of the standard of reference may be taken as the unit of measure for a given quantity; a given length may be measured by 1 foot, or by 1000 feet, or by $\frac{1}{1000}$ of a foot. But in beginning the explicit treatment of fractions it is better to use certain standard measures, their subdivisions, and their relations to one another. Thus, as a process of analysis-synthesis, the foot is defined by $\frac{12}{12}$, the yard

by $\frac{3}{3}$, the pound by $\frac{16}{16}$, the dollar by $\frac{10}{10}$ or by $\frac{100}{100}$; where 12 refers to inches, 3 to feet, 16 to ounces, 10 to dimes, and 100 to cents. Familiarity with fractions thus defined by and connected with the ordinary scales of measurement means easy mastery of all forms of fractions as a mode of definite measurement.

3. As to the teaching of fractions, it will be enough, for the present, to note the following points:

1. In the formal treatment of fractions nothing new is involved; there is simply a *conscious* direction of attention to ideas and processes which, under right teaching, have been used from the first in the formation of numerical ideas, and which have been further developed in the fundamental arithmetical operations.

2. As in "integers" so in teaching fractions, the idea and process of measurement should be ever present. To begin the teaching of fractions with vague and undefined "units" obtained by breaking up equally undefined wholes—the apple, the orange, the piece of paper, the pie—may be justly termed an irrational procedure. Half a pie, e. g., is not a numeral expression at all, unless the pie is defined by weight or volume; the constituent factors of a fraction are not present; the unity of arithmetic is ignored; the process of fractions is assumed to be something different from that of number as measurement; it becomes a question—it actually has been questioned—whether a fraction is really a number; and all this in spite of the fact that from the beginning fractions are implicit in all operations; that from first to last the process of number as a psychical act is a process of fractions.

3. The primary step in the explicit teaching of frac-

NATURE OF DIVISION AND FRACTIONS. 141

tions—that is, in making the habit of fractioning already formed an object of analytical attention—is to make perfectly definite the child's acquaintance with certain standard measures, their subdivisions and relations. In all fractions—because in all exact measurement—there must be a *definite* unit of measure. This implies two things: (*a*) The definition of a standard of reference (the "primary" unit) in terms of its own unit of measure; (*b*) the measurement of the given quantity by means of this "derived" unit. If the foot is unit of measure, it is unmeaning in itself; it must be mastered, must be given significance by relating it to other units in the scale of length; it is 1 (yard) ÷ 3 in one direction; or (taking the usual divisions of the scale) it is $\frac{12}{12}$ (i. e., $\frac{1}{12} \times 12$) in the other direction, i. e., as measured in inches. The teaching of fractions, then, should be based on the ordinary standard scales of measurement; on the fundamental process of parting and wholing in measurement, and not upon the qualitative parts of an undefined unity.

4. Under proper teaching of number as measurement the pupil soon learns to identify instantly 4 inches, $\frac{4}{12}$ foot, $\frac{1}{3}$ foot as expressions for the same measured quantity. He is led easily to the conscious recognition of the true meaning of fractions as a means of indicating the exact measurement of a quantity in terms of a measuring unit which is itself exactly measured.

5. Addition and subtraction of fractions involve the principle of ratio, multiplication and division the principle of proportion. In all cases the meaning of fractions as denoting definitely measured quantity should be made clear. For example, not $\frac{1}{2} \times \frac{3}{4}$, but $\frac{1}{2}$ foot $\times \frac{3}{4}$;

not $\frac{3}{4} \times \frac{7}{8}$, but $\$\frac{3}{4} \times \frac{7}{8}$, as indicating, e. g., the cost of $\frac{7}{8}$ yard of cloth at $\$\frac{3}{4}$ a yard.

Since a fraction expresses a quantity in a form for comparison with other quantities of the same kind, the fundamental operations as applied in fractions carry out these comparisons. Addition is always (as in "whole numbers") of homogeneous quantities—i. e., those measured in terms of some unit of length, surface, volume, time, etc.; so with subtraction. All the first examples should deal only with definite measures; after the principle is quite familiar, and only then, fractions having denominators not corresponding to any existing scale of measurement—e. g., 17, 49. 131—may be introduced for the sake of securing mechanical facility.

The same remark applies to multiplication and division of fractions—operations which involve no principles different from the corresponding operations with "whole numbers." Multiplication of fractions is *multiplication*, and division is *division;* they are not new processes under old names. They make explicit use of ratio (the comparison of quantities), which is implied in the operations with "integers," by defining the measuring unit which defines a measured quantity. They put in shorthand, as it were, the complete psychical process of measurement, and thus make a severer demand on conscious attention. But if number has been from the first taught upon the psychological method, the pupil will be quite prepared to meet this demand. There will be nothing strange in reducing fractions to a common denominator, nor any mystery in a product less than the multiplicand, or in a quotient greater than the dividend; so far as the nature of the processes is con-

NATURE OF DIVISION AND FRACTIONS.   143

cerned, $\$\tfrac{3}{4} + \$\tfrac{4}{5}$ will be just as intelligible as $\$3 + \$4$. If, too, the nature and relation of *times* and measuring *parts* have become familiar, there will be no more mystery in 18 feet $\times \tfrac{1}{2} = 9$ feet than in measuring half-way across a room 18 feet wide; the peculiar thing would be if taking a quantity only a part of a time did *not* give a smaller quantity.

So in division, when the mutual relation between times and parts is understood, the operation $\$\tfrac{4}{5} \div \$\tfrac{1}{10}$, or $\$\tfrac{4}{5} \div \tfrac{1}{10}$, is just as intelligible as $\$80 \div \$10$, or as $\$80 \div 10$. To say that the quotient *eight*, the result of $\$\tfrac{4}{5} \div \$\tfrac{1}{10}$, is greater than the dividend ($\$\tfrac{4}{5}$) is to talk nonsense; is to compare incomparable things—is to confuse parts with *times*, quantity with number, matter with a psychical process.

CHAPTER VIII.

ON PRIMARY NUMBER TEACHING.

The Number Instinct.—We have seen that number is not something impressed upon the mind by external energies, or given in the mere perception of things, but is a product of the mind's action in the measurement of quantity—that is, in making a vague whole definite. Since this action is the fundamental psychical activity directed upon quantitative relations, the process of numbering should be attended with interest; that is, contrary to the commonly received opinion, the study of arithmetic should be as interesting to the learner as that of any other subject in the curriculum. The training of observation and perception in dealing with nature studies is said to be universally interesting. This is no doubt true, as there is a hunger of the senses—of sight, hearing, touch—which, when gratified by the presentation of sense materials, affords satisfaction to the self. But we may surely say with equal truth that the exercise of the higher energy which works upon these raw materials is attended with at least equal pleasure. The natural action of attention and judgment working upon the sense-facts must be accompanied with as deep and vivid an interest as the normal action of the observing powers through which the sense-facts are acquired.

For numerical ideas involve the simplest forms of this higher process of mental elaboration; they enter into all human activity; they are essential to the proper interpretation of the physical world; they are a necessary condition of man's emancipation from the merely sensuous; they are a powerful instrument in his reaction against his environment; in a word, number and numerical ideas are an indispensable condition of the development of the individual and the progress of the race. It would therefore seem to be contrary to the "beautiful economy of Nature" if the mind had to be *forced* to the acquisition of that knowledge and power which are essential to individual and racial development; in other words, if the conditions of progress involved other conditions which tended to retard progress.

The position here taken on theoretical grounds, that the normal activity of the mind in constructing number is full of interest, is confirmed by actual experience and observation of the facts in child life. There are but few children who do not at first delight in number. Counting (the fundamental process of arithmetic) is a thing of joy to them. It is the promise and potency of higher things. The one, two, three of the " six-years' darling of a pygmy size " is the expression of a higher energy struggling for complete utterance. It is a proof of his gradual emergence from a merely sensuous state to that higher stage in which he begins to assert his mastery over the physical world. We have seen a first-year *class*—the whole class—just out of the kindergarten, become so thoroughly interested in arithmetic under a sympathetic and competent teacher, as to *prefer* an exercise in arithmetic to a kindergarten song or a romp in the playground.

Arrested Development.—Since, then, the natural action of the child's mind in gaining his first ideas of number is attended with interest, it seems clear that when under the formal teaching of number that interest, instead of being quickened and strengthened, actually dies out, the method of teaching must be seriously at fault. The method must lack the essentials of true method. It does not stimulate and co-operate with the rhythmic movement of the mind, but rather impedes and probably distorts it. The natural instinct of number, which is present in every one, is not guided by proper methods till effective development is reached. The native aptitude for number is continually baffled, and an artificial activity, opposed to all rational development of numerical ideas, is forced upon the mind. From this irrational process an arrested development of the number function ensues. An actual distaste for number is created; the child is adjudged to have no interest in number and no taste for mathematics; and to nature is ascribed an incapacity which is solely due to irrational instruction. It is perhaps not too much to say that nine tenths of those who dislike arithmetic, or who at least feel that they have no aptitude for mathematics, owe this misfortune to wrong teaching at first; to a method which, instead of working in harmony with the number instinct and so making every stage of development a preparation for the next, actually thwarts the natural movement of the mind, and substitutes for its spontaneous and free activity a forced and mechanical action accompanied with no vital interest, and leading neither to acquired knowledge nor developed power.

Characteristics of this defective method have been frequently pointed out in the preceding pages, and it is unnecessary to notice them here further than to caution the teacher against a few of them, which it is especially necessary to avoid.

Avoid what has been called the "fixed-unit" method. No greater mistake can be made than to begin with a single thing and to proceed by aggregating such independent wholes. The method works by fixed and isolated unities towards an undefined limit; that is, it attempts to develop accurate ideas of quantity without the presence of that which is the essence of quantity— namely, the idea of *limit*. It does not promote, but actually warps, the natural action of the mind in its construction of number; it leaves the fundamental numerical operations meaningless, and fractions a frowning hill of difficulty. No amount of questioning upon one thing in the vain attempt to develop the idea of "one," no amount of drill on two such things or three such things, no amount of artificial analysis on the numbers from one to five, can make good the ineradicable defects of a beginning which actually obstructs the primary mental functions, and all but stifles the number instinct.

Avoid, then, excessive analysis, the necessary consequence of this "rigid unit" method. This analysis, making appeals to an undeveloped power of numerical abstraction, becomes as dull and mechanical and quite as mischievous in its effects as the "figure system," which is considered but little better than a mere jugglery with number symbols.

Avoid the error of assuming that there are exact numerical ideas in the mind as the result of a number of things before the senses. This ignores the fact that number is not a thing, not a property nor a perception of things, but the result of the mind's action in dealing with quantity. Avoid treating numbers as a series of separate and independent entities, each of which is to be thoroughly mastered before the next is taken up. Too much thoroughness in primary number work is as harmful as too little thoroughness in advanced work.

Avoid on the one hand the simultaneous teaching of the fundamental operations, and on the other hand the teaching which fails to recognise their logical and psychological connection.

Avoid the error which makes the "how many" alone constitute number, and leaves out of account the other co-ordinate factor, "how much." The *measuring* idea must always be prominent in developing number and numerical operations. Without this idea of measurement no clear conception of number can be developed, and the real meaning of the various operations as simply phases in the development of the measuring idea will never be grasped.

Avoid the fallacy of assuming that the child, to know a *number*, must be able to picture all the numbered units that make up a given quantity.

Avoid the interest-killing monotony of the Grube grind on the three hundred and odd combinations of half a dozen numbers, which thus substitutes sheer mechanical action for the spontaneous activity that simultaneously develops numerical ideas and the power to retain them.

Rational Method.—The defects which have been enumerated as marking the "fixed-unit" method suggest the chief features of the psychological or rational method. This method pursues a diametrically opposite course. It does not introduce one object, then another "closely observed" object, and so on, multiplying interesting questions in the attempt to develop the number one from an accurate observation of a single object. It does as Nature prompts the child to do: it begins with a quantity—a group of things which may be measured—and makes school instruction a continuation of the process by which the child has already acquired vague numerical ideas. Under Nature's teaching the child does not attempt to develop the number one by close observation of a single thing, for this observation, however close, will not yield the number one. He develops the idea of one, and all other numerical ideas, through the measuring activity; he counts, and thus measures, apples, oranges, bananas, marbles, and any other things in which he feels some interest. Nature does not set him upon an impossible task—i. e., the getting of an idea under conditions which preclude its acquisition. She does not demand numerical abstraction and generalization when there is nothing before him for this activity to work upon. Let the actual work of the schoolroom, therefore, be consistent with the method under which by Nature's teaching the child has already secured some development of the number activity.

In all psychical activity every stage in the development of an instinct prepares the way for the next stage. The child's number instinct begins to show itself in its working upon continuous quantity—that is, a whole requiring measurement. Every successive step in the entire course of development should harmonize with this initial stage. To get exact ideas of quantity the mind must follow Nature's established law; must measure quantity; must break it into parts and unify the parts, till it recognises the one as many and the many as one. There can be no possible numerical abstraction and generalization without a quantity to be measured. Where, then, does the "single closely observed object" come in as material for this parting and wholing?

Beginning with a group is in harmony with Nature's method; promotes the normal action of the mind; gives the craving numerical instinct something to work upon, and wisely guides it to its richest development. This psychological method promotes the natural exercise of mental function; leads gradually but with ease and certainty to true ideas of number; secures recognition of the unity of the arithmetical operations; gives clear conceptions of the nature of these operations as successive steps in the process of measurement; minimizes the difficulty with which multiplication and division have hitherto been attended; and helps the child to recognise in the dreaded *terra incognita* of fractions a pleasant and familiar land.

Forming the Habit of Parting and Wholing.— The teacher should from the first keep in view the importance of forming the *habit* of parting and wholing. This is the fundamental psychical activity; its goal is to grasp clearly and definitely by one act of mind a whole of many and defined parts. This primary activity working upon quantity in the process of measurement gives rise to numerical relations; the incoherent whole is made definite and unified—becomes the conception of a unity composed of units. Every right exercise of this activity gives new knowledge and an increase of analytic power. At last the *habit* of numerical analysis is formed, and when it is found requisite to deal with quantity and quantitative relations, the mind always conceives of quantity as made up of parts—measuring units; not invariable units, but units chosen at pleasure or convenience; parts, given by the necessary activity of analysis, a whole from the parts by the necessary activity of synthesis. This means that always and inevitably from first to last the process of fractioning is present.

A Constructive Process.—This wholing and parting, as far as possible, should be a constructive act. The physical acts of separating a whole into parts and reuniting the parts into a whole lead gradually to the corresponding mental process of number: division of a whole into exact parts, and the reconstruction of the parts to form a whole. It can not be said that even the physical acts are wholly mindless, for even in these acts there must be at least a vague mental awareness of the relation of the parts to one another and to the whole. These physical acts of wholing and parting under wise direction lead quickly, and with the least expenditure of energy, to clear and definite percepts of related things, and finally to definite conceptions of number. The child should be required to exercise his activity, to do as much as possible in the process, and to notice and state what he is really doing. He should actually apply, for instance, the measuring unit to the measured quantity. If the foot is measured by two 6-inch or three 4-inch or four 3-inch units, let him first apply the number of actual units—two 6-inch, three 4-inch, and four 3-inch units—to make up the foot, and so on. By using the actual number of parts required he will have a more definite idea of the construction of a whole than if he simply applies one of the measuring units the necessary number of times. This operation with the actual units should precede the operation by which the whole is mentally constructed by applying or repeating the single unit of measurement the required number of times. It is the more concrete process, and is an effective exercise for the gradual growth of the more abstract times or ratio idea.

When the child actually uses the 1-inch or the 3-inch unit to measure the foot, his ideas of these units as well as of the measured whole are enlarged and defined. He applies the inch to measure the foot, and this to measure the yard, and the yard to measure the length of the room and other quantities. Let him freely practise this constructive activity, thus practically applying the psychological law, "Know by doing, and do by knowing." The 2-inch square is separated into four inch squares, or sixteen half-inch squares, and these measuring units are put together again to form the whole. Similarly a rectangle 2 inches by 3 inches, for example, is divided into its constituent inch squares or half-inch squares, and again reconstructed from the parts. A square is divided into four right-angled isosceles triangles, into eight smaller triangles, and the parts rhythmically put together again.

Value of Kindergarten Constructions.—In this connection it may be noted that most of the exercises of the kindergarten can be effectively used for training in number. The constructive exercises which are so prominent a feature in the kindergarten are admirably adapted to lead gradually to mathematical abstraction and generalization. No doubt much has been done in this direction, but much more could be done were the teacher versed in the psychological method of dealing with number. No one questions the general value of kindergarten training, which on the whole is founded on sound psychological principles; but, on the other hand, no educational psychologist doubts that its philosophy as commonly understood needs revision, and that its methods are capable of improvement. If its aim is, as it should be, an effective preparation of the child for his subsequent educational course, it is thought that its practical results are far from what they ought to be. It is often maintained with considerable force that kindergarten methods should be introduced into the primary and even higher schools. On the other hand, something might be said with a good show of reason in favour of introducing primary and grammar school methods into the kindergarten. What is radically sound in the kindergarten methods will harmonize with what is radically sound in the methods of the public school. On the other hand, what is psychologically sound in the methods of the public school should at least influence the aims and methods of the kindergarten. Is the present kindergarten training, speaking generally, really the best preparation for the training given in a thoroughly good public school? The function of such a school is to give the best possible preparation for life by means of studies and discipline, which, as far as inevitable limitations permit, secure at the same time the best possible development of character. Among these studies the three R's must always hold a prominent place, in spite of theories which seem to assume that language, the complement of man's reason, and number, the instrument of man's interpretation and mastery of the physical world, are not essential to human advancement, and may therefore be degraded from the central position which they have long occupied to one in which they are the subjects of merely haphazard and disconnected teaching.

The practical methods founded on these theories seem to treat the world of Nature as one whole, which even the child may grasp in its infinite diversity and total unity. The "flower in the crannied wall" is made the central point around which all that is knowable is to be collected. But as the human mind is limited, and must move obedient to the law of its constitution, the theories and methods which overlook these facts are not likely permanently to prevail; and the old subjects that have stood the test of time will no doubt stand the test of the most searching psychological investigation, and regain their full recognition as the "core" subjects of the school curriculum.

Does the kindergarten, then, accomplish all that may be done as a preparation for such a curriculum? It is to be feared that with regard to many of them the answer must be in the negative; and this is perhaps especially true concerning the subject of arithmetic. We have known the seven-year-old "head boy" of a kindergarten, conducted by a noted kindergarten teacher, who could not recognise a quantity of three things without counting them by ones. Being asked to begin the construction of a certain form at a distance of three inches from the edge of the table, he invariably had to count off carefully the inch spaces—a clear proof, it is thought, that his training had not been the best possible preparation for the arithmetical instruction of the higher schools.

It is certain that arithmetical instruction in the higher grade of school may be greatly improved; it is alike certain that as a preparation for this better instruction the training of the kindergarten also may be greatly improved; and there is every reason to believe that with this improvement its rational training in other things, its ethical aim, its educative interest, and its character-forming spirit, would be materially enhanced. In some kindergartens, at least, the monotony of continuous play would be pleasurably relieved by a little recreation at work. It is certain that the interest associated with many exercises necessarily connected with the number activity, especially the constructive and analytic processes of the kindergarten, can be made under right teaching the means by which numerical ideas may be gradually and pleasantly worked out. The little builder of many forms of beauty and utility would in due time find, when the inevitable and harder tasks began, that he had been building better than he knew. There is surely something lacking either in the kindergarten as a preparation for the primary school, or in the primary school as a continuation of the kindergarten, when a child after full training in the kindergarten, together with two years' work in the primary school, is considered able to undertake nothing beyond the "number 20." It might reasonably be maintained that, under rational and therefore pleasurable training of the number instinct in the kindergarten, the child ought to be arithmetically strong enough to make immediate acquaintance with the number 20, and rapidly acquire—if he has not already acquired—a working conception of much larger numbers.

Important Points of Rational Method.—In applying the rational method of teaching arithmetic there are important things that the teacher must keep in view if he is to aid the child's mind to work freely and naturally in the evolution of number. The child's mind must be guided along the lines of least resistance to the *true idea of number*. This movement, in the very nature of things, must be slow as compared with the gathering of sense facts; but under the psychological method it may be sure and pleasurable. The result aimed at can not be reached by banishing the word *times* from arithmetic; nor by working continually with indefinite units of measure; nor by exclusive attention to manual occupations under the vague idea that physical separation of things is analysis of thought; nor by making counting—emphasizing the vague *how many*—the single purpose, and unmeasured units the sole matter of the exercises, to the exclusion of the *how much*, and the measuring idea which is the essence of number; nor by substituting for the rhythmic and spontaneous action of the child's mind in dealing with wholes, both qualitative and quantitative, a minute and formal analysis which properly finds place only in a riper stage of mental growth; nor by any amount of drill, however industrious and deviceful the drill-master, which substitutes mechanical action and factitious interest for spontaneous action and intrinsic interest, the very life of the self-developing soul.

Number is the measurement of quantity, and therefore the only solid basis of method and sure guide for the teacher is the *measuring* idea.

1. *The Measured Whole.*—The factors in number are, as before shown, the unity (the whole of quantity) to be measured, the unit of measurement, and the times of its repetition—the number in the strictly mathematical and psychological sense of the word. The teacher must bear in mind the distinction between unity and unit as fundamental. The entire difference between a good method and a bad method lies here, because the essential principle of number lies here. Vague unity, units, defined unity, is the sequence as determined by psychological law. In the child's first dealing with number there must be the group of things, the whole of quantity to start from; and in every step of the initial stage the idea of a whole to be measured is to be kept prominent. In addition, there is a whole (the sum) to be made more definite by putting together its component parts (addends)—not equal *measuring* units, but each part defined by a common unit—so as to completely define the quantity in terms of this specifically defined unit. In subtraction we have a given quantity (minuend) and a component part (subtrahend) of it to find the other component (remainder)—a process which helps to a more definite idea of the given whole, and especially makes explicit the vague idea of the "remainder" with which we began. In multiplication there is

given a quantity (multiplicand) defined by a measuring unit and the times (multiplier) of its repetition; and the process makes the quantity articulately defined (in the product) by substituting a more familiar unit for the derived unit of measurement; in other words, by expressing the quantity in terms of the primary unit by which the derived unit itself is measured. In division we have a whole quantity given (dividend), and one of two related measuring parts (the divisor) to find the other part (quotient), and the operation makes clearer the whole magnitude, and at the same time makes the first vague idea of the other measuring part (quotient) perfectly definite. Briefly, in all numerical operations there is some magnitude to be definitely determined in numerical terms, and the arithmetical operations are simply related steps expressing the corresponding stages of the mental movement by which the vague whole is made definite. Keep clearly in mind, therefore, the inclusive magnitude from which and within which the mental movement takes place—which justifies and gives meaning to both the psychical process and the arithmetical operation.

2. *The Unit of Measure—Its True Function.*— From the vague unity, through the *units*, to the definite unity, the *sum*, is the law of mental movement. The second point of essential importance is to make clear the idea of the unit of measure. More than half the difficulty of the teacher in teaching, and the learner in learning, is due to misconception of what the "unit" really is. It is not a single unmeasured object; it is not even a single defined or measured thing; it is any *measuring part* by which a quantity is numerically de-

fined; it is (in the crude stage of measurement) one of the *like* things *used* to measure a collection of the things; it is (in the second or exact stage of measurement) one of the *equal* parts used to measure an exactly measured quantity. It is *one* of a necessarily *related* many constituting a whole.

It is, therefore, an utterly false method to begin with an isolated object—false to the fact of measurement, false to the free activity of the mind in the measuring process. Nor is the defect to be remedied by introducing another isolated object, then another, and so on. The idea of a unit can begin only from analysis of a whole; it is completed only by relating the part to the whole, so that it is finally conceived at once in its isolation and in its unity in the whole. Not only do we not begin with a single object and "develop one," but also even in beginning, as psychology demands, with a group of objects, we are not to begin with the single object to measure the quantity—at least we are not to emphasize the single object as pre-eminently the measuring unit. We separate twelve beans, for example, not into 12 parts, but into 2 parts, then 3 parts, etc.; that is, we measure by 6 beans, by 4 beans, by 3 beans, by 2 beans; and the resulting *numbers* for the one measured quantity are two, three, four, six; and each of the measuring parts to which the numbers are applied is a *unit*, is ONE. So in building up the measured foot with 6-inch, 4-inch, 3-inch, 2-inch measures—each in turn measured off in inches—there are two *ones*, three *ones*, four *ones*, six *ones*. The point to be kept in view is to prevent the mischievous error of regarding the unit as a single OBJECT, a fixed qualitative or an indivisible quantitative unity, instead of simply a measuring part—a means of measuring a magnitude.

The Unit itself Measured.—As necessary to the growth of the true conception of unit as a measuring part, the idea of the unit as a unity of measured parts must be clearly brought out. The given quantity is measured by a certain unit; this unit itself is a quantity, and so is made up of measuring parts. This idea must be used from the beginning; it is absolutely essential to the clear idea of the unit, and of number as measurement of quantity. Beginning with a group of 12 objects requiring measurement, or with counters representing such objects, we have them counted off into two equal parts, noting the relation of the parts to one another and to the whole; then each of these *two* units (half of the given whole) is counted off into two equal parts, and the relation of these minor parts to each other and to the whole they compose is noticed; then each of the first units of measurement (halves of the given whole) is counted off into three equal parts, and their relation to one another and to the whole which *they* make is carefully observed; and so on, with similar exercises in parting and wholing. Such constructive exercises *help* in the growth of the true idea of the unit as a measuring part, which is or may be itself measured by other units. But the true idea of the essential property of the unit—its *measuring function*—can be fully developed only by exercises belonging to the second stage of measurement, in which exact and equal units are used for precise measurement. These measurements of groups of *like* things (apples, oranges, etc.) by groups which are themselves measured by still smaller groups, *must* be supplemented by the use of exactly measured quantities—quantities defined by *equal* units, which in turn are measured by other *equal* units. Without such exercises there can be no adequate conception of measurement, or of number as the tool of measurement, or of the real meaning of multiplication and division, and especially of fractions, the full and precise statement of the measuring process. To free arithmetic from the tyranny of irrational method, an indispensable step is the emancipation of the unit from the cast-iron fetters which have paralyzed its *measuring* function.

3. *The Idea of Times.*—With the intelligent use of these constructive exercises to make clear the idea of the unit, there is necessarily growth towards recognition of the times of repetition of the unit to make up some magnitude—towards, that is, the true idea of number. To discuss the evolution of this idea would be to repeat in the main what has been laid down in the preceding paragraphs. It is therefore necessary only to state explicitly the chief things to be considered as bearing upon the natural growth of the idea of times —i. e., of *number* in the strict sense of the word.

(*a*) The preliminary operations as already illustrated —dealing with groups of like things, and so leading to a working idea of the unit as *measuring* part—are to be supplemented by constructive acts with exactly measured quantities. The measured whole must be analyzed into its measured units, and again built up from these parts. For example, exercises such as the quantity 12 apples measured by the unit 4 apples, by the unit 3 apples, etc., must be supplemented by exercises such as the quantity 12 inches measured by the unit 4 inches, by the unit 3 inches, etc.; or the quantity 20 cents measured by the unit 10 cents, by the unit 5 cents, etc. The movement towards the real number idea began in operations with undefined units, and is strengthened by these supplementary exercises with exactly measured quantity; there is a more rapid growth towards the numerical discriminating and unifying power. (*b*) Count by *ones*, but not necessarily by single things; in fact, to avoid the fixed unit error, do not begin with counting single things. The 12 things in the group have been measured off, for example, into four groups, or into three groups; these are units, are *ones*, and in counting there is a first *one*, a second *one*, a third *one*—that is, in all "three *times*" one; and so with the four *ones* when the quantity is divided into four equal parts. Proceed similarly with exactly measured quantities: the four 3-inch *ones* or the six 2-inch *ones* making up the linear foot, or other exactly measured quantity. As before said, the child first of all sees related things, and with the repetition of the exercises—parting and wholing—begins to feel the relations of things, and in due time consciously recognises these relations, and the goal is at last reached—a definite idea of number.

(*c*) *Use the Actual Units.*—In these constructive processes let the child at first use—as before suggested—the actual concrete units to make up or equal the measured quantity; then apply the *single* concrete unit the requisite number of "times." In the first case, in measuring, for example, a length of 12 feet, four actual units of measure (3 feet) are put together to equal the 12 feet; in the second case, *one* unit is applied, laid down, and taken up three *times*. This application of the single unit so many times is an important step in the process of numerical abstraction and generalization; it is from the less abstract and more concrete to the more abstract and less concrete. It may be noted, also, that the other senses, especially the sense of hearing, may be made to co-operate with sight in the evolution of the *times* idea. Appeal by a variety of examples to the trusty eye, but appeal also to the trusty ear—strokes on a bell, taps on the desk, uttered syllables, etc. Here, as in all other cases, we do not confine ourselves to single bell-strokes or syllables; we count the number of double strokes, triple strokes; of double and triple syllables, as, for example, *oh, oh; oh, oh; oh, oh*—i. e., 3 counts of two sounds each, etc.

Counting and Measuring.—In the separating and combining processes referred to, counting goes on. This is at first chiefly mechanical, and care must be taken in the interest of the number idea to make it become rational. Through practice in parting and wholing the idea of the function of the unit is gradually formed; it is the concrete, spatial thing used to measure quantity. The point is, not to neglect either the spatial element or the other essential factor in number, the counting, the actual relating process.

In the method of number teaching usually followed, counting is the prominent thing, to the almost total exclusion of the *measuring* idea; the emphasis is upon the how many, with but little attention to the how much. But the counting is largely mechanical. There is a repetition of names without definite meaning. The child is groping his way towards the light. He can not help feeling, as he counts his units, that one, two, three is not so much—because it is not so many—as one, two, three, four. These first vague ideas must be made clear and definite; the natural movement of the mind is aided by the proper presentation of right material; the initial mechanical operation of naming the units in order gives place to an intelligent relating of the units to one another, and finally to a conscious grasp of the relation of each to the unified whole; the counting—one, two, three, etc.—is now a rational process.

So much; so many.—In the development of this rational process there must be no divorce between the how much and the how many, between the measuring process and the results of measurement. The so much is determined only by the so many, and the so many has significance only from its relation to the so much. These are co-ordinate factors of the idea of number as measurement. Now, the development of counting—determining the how many that defines the how much—is aided by symmetrical arrangements of the units of measure (see page 34). The child at first counts the units one, two, . . . six, with only the faintest idea of the relations of the units in the numbers named. Both the analytic and relating activities are greatly aided by the rhythmic grouping of the units of measure, or of the counters used to represent them; the mastery of the number relations (of both addition and multiplication) as so many units making up a quantity, becomes much easier and more complete. Thus, when exercises in parting and wholing (accompanied with counting) a quantity, say a length of 12 inches, have given rise to even imperfect ideas of unit of measurement and times of repetition, the symmetric forms may be used with great advantage; indeed, they may be used in the exercises from the first. We have counted, according to the unit of measure used, one part, two parts; one part, two parts, three parts, etc. Both the times and the unit values are more easily grasped through the number forms; for example, six, one of the two measuring units, may be shown as a whole of related units (threes, twos, ones) as in the arrangement, ❘ ❘ ❘ ; and so on with the whole quantity and all its minor parts (addition) and repeated units. Real meaning is given to the operation of counting when, instead of using unarranged units, we have the rhythmic arrangement:

∷ | ∷ | ∷ | ∴ | ∶ | •

Six,   five,   four,   three,   two,   one.

The actual values of the measuring units, and the meaning of counting—necessarily related processes—are fully brought out. Six is at last perceived as six without the necessity of counting.

CHAPTER IX.

ON PRIMARY NUMBER TEACHING.

Relation between Times and Parts.—With the growth of the idea of the unit as a measure itself measured by minor units, and of number as indicating times of repetition of a unit of measure, there is gradually developed a clear idea of the relation between the value of the actual measuring unit (as made up of minor units) and the number of them in the given quantity; in other words, of the relation between the number of derived units in the quantity and the number of primary units in the derived unit. A common error, as has been often pointed out, is that of making too broad a distinction between these related factors in the measuring process. They are said to be totally different conceptions. It has been shown that they are absolutely inseparable. They are, in any and every case, two aspects of the same measurement. The direct unit in a given measurement is not wholly concrete; it is a quantity measured by a number of *other* units; and so it involves, as every measured quantity involves, the space element in the single concrete (minor) unit, and the abstract element in the number applied to the unit. When we speak of the "size" of the numbered parts (derived units) composing a given quantity, we mean the number of minor units of which one part is composed. When, for instance, we conceive of $15 as measured by the unit $3, we get the number five; when we are required to divide $15 into five equal parts, we are searching for the "size" ($3) of the measuring unit—i. e., for the numerical value of the unit in terms of the minor unit ($1) by which it is measured.

The relation, then, between "times and parts" is the relation between the number of derived units in the measured quantity and the number of primary units in the derived unit. It is clear that the rational processes of parting and wholing that ultimately give clear ideas of *unit* and number, must also bring out clearly the relation between these two factors in measured quantity: the smaller the unit the larger the number; or, the number of the measuring units in the quantity varies inversely as the number of primary units in the derived unit. Measuring a length of one foot by a 6-inch unit, by a 3-inch unit, and by a 1-inch unit, the numbers are respectively 2, 4, and 12; measuring a length of one decimetre by 10, 20, 30, 40, 50 centimetres, the numbers are respectively 10, 5, $3\frac{1}{3}$, $2\frac{1}{2}$, 2; measuring $20 by the $1 unit, $2 unit, $4 unit, the numbers are 20, 10, 5, etc. In the constructive exercises already described, attention to measuring unit and its times of repetition must lead to the conscious recognition of this principle, which is fundamental in number as measurement. It has already been given in the complete statement in "fractional" form of the process of measurement: *Any measured quantity* may be expressed in the form $\frac{1}{1}, \frac{2}{2}, \frac{3}{3}, \frac{4}{4} \ldots \frac{n}{n}$. This principle is, of course, the basal principle in the treatment of fractions (because it is the primary principle of number, and "fractions" are numbers)—namely, both terms of a fraction may be multiplied or divided by the same number without altering the value of the fraction.

The Law of Commutation.—The development of this principle through the rational use of its idea—that is through the use of the facts supplied by sense-perception in the rational use of things—is the development of the psychological law of commutation which is primary and essential in all mathematics. The method that ignores this necessary relation between times and parts, or regards them as totally different things, never leads to a clear conception of this important principle. It is, as a consequence, always finding difficulties where, for rational method, none really exist. The principle is difficult for the child only when the method is wrong. With right presentation of material he can have no difficulty in seeing that the larger the units the smaller their number in a given quantity. When he counts out a collection of 24 objects into piles of 3 each, and into piles of 6 each, can he fail to see that the respective numbers differ? And with rightly directed attention to the concrete processes, may he not be led slowly, perhaps, but surely, to a clear thought of how and why they differ? The learner can not help seeing, for example, the difference between the 2-inch unit and the 6-inch unit, and the corresponding difference between six times and three times. To see clearly is to think clearly; there is a rationality in rationally presented facts, and this rationality leads with certainty to a complete recognition of the meaning—the law—of the facts.

Should be Used from the First.—Thus the idea of the law of commutation can be used from the first; rather, must be used if clear and adequate ideas of number are to be gained, and numerical operations are to be a thing of meaning and of interest. As already seen, it is impossible to *know* 12 objects, as measured by *fours*, without at the same time knowing it as measured by *threes*. So it is in the actual operation with things: we count out a quantity of 12 things into groups of *four* things each, and find the number of the groups to be *three;* and we count out the 12 things into four groups, and find in each group *three* things. In both cases the operations are alike; in neither is it possible to get the result without using counts of *four* things each. The child counts out 12 things into groups of 4 things each; how many groups? He counts them out into 4 groups; how many things in each group? In both cases he sees that he counts by fours. He counts 10 things in groups of 5 things each; how many groups? He counts them out into 5 groups; how many in each group? In both operations he sees that he counts out into groups of five things each. Twenty cents are to be equally divided among 5 boys; how many cents will each boy get? Twenty cents are divided equally among a number of boys, giving each boy 5 cents; how many boys are there? The child performs the operation with counters, and finds in both cases *four*, meaning 4 boys in the one case, 4 cents in the other; he sees that in both examples the operation was a counting by *fives*, and he will soon be in possession of the important truth—which many teachers, and even teachers of teachers, seem not to know—that one *fifth* of 20 units of any kind is *four* units, *because* *five* units must be repeated *four* times to measure 20 units.

There will be but little difficulty in further illustrating the principle by applying it to exact measurement. Three 4-inch measures make up the linear foot; the child sees that one inch counted out of each of the 3 4-inch units gives a whole of 3 inches; another inch thus counted out gives another whole of 3 inches, etc.; so that in all there are 4 units of 3 inches each. Since, however, the principle is more distinctly illustrated by symmetrical groupings of the measuring units (see page 34), the separate primary units may be used with advantage—for example, for the 4-inch measure use the four single units which compose it. With the proper use of such parting and wholing exercises the child can not fail to comprehend in due time this fundamental principle in number and numerical operations. Seeing clearly that the unit is composed of minor units, that it is repeated to make up or measure some quantity, he can not fail to see that each and all of its minor units are repeated the same number of times.

What has been joined together by a psychical law can not be divorced without checking or distorting mental growth. It is known from experience that when the constructive exercises already referred to are carried on under rational direction, the use of the related ideas grows naturally and surely into a conscious recognition of the relations: how much and how many, quantity and its instrument of measurement, parts and times, measuring unit and measured whole, measuring minor unit and its measured minor whole, correlated factors of product, correlated processes of division, are all seen in their true logical and psychological interrelation. These things are organically connected by necessary laws of thought; the method which is rationalized by this idea makes arithmetic a delight to the pupil and a powerful educating instrument. A method which violates this necessary law of mind in dealing with quantity—constantly obstructing the original action of the mind—makes arithmetic a thing of rule and routine, uninteresting to the teacher, and probably detested by the learner.

Make Haste slowly.—As already suggested, time is necessary for the completion of the idea of number. Under sound instruction a *working* conception suitable for a primary stage of development may be readily acquired, and may be used for higher development. But a perfectly clear and definite conception of number is a product of growth by slow degrees. The power of numerical abstraction and generalization can not be imparted at will even by the most painstaking teacher. Hence the absurdity of making minute mechanical analysis a substitute for nature's sure but patient way. It seems to be thought that mechanical drill upon a few numbers—a drill which, if rational, would really use ratio and proportion—will in some unexplained way "impart" the idea of number to the child apart from the self-activity of which alone it is the product. And this fallacious idea is strengthened by the fluent chatter of the child—the apt repeater of mere sense facts—about the "equal numbers in a number," the "equal numbers that make a number," and all the rest of it. This routine analysis and parrotlike expression of it are a direct violation of the psychology of number. The idea of number can not be got from this forcing process. The conscious grasp of the idea, we repeat, must come from rational *use* of the idea, and is all but impossible by monotonous analyses with a few "simple" numbers; it is absolutely impossible in the immature state of the child-mind in which it is attempted. The child must, once more, freely and rationally *use* the ideas; must operate with many things, using many numbers, before the idea of number can possibly be developed. We are omitting the things the child can do—rationally *use* numerical ideas—and forcing upon him things that he can not do—form at once a complete conception of number and numerical relations. It is high time to change all this: to omit the things he can not do, and interest him in the things he can do. In the comparatively formal and mechanical stage there must be a certain amount of mechanical drill—mechanical, yet in no small degree disciplinary, because it works with ideas which, though imperfect, are adequate to the stage of development attained, and through rational use become in due time accurate scientific conceptions. Besides valuable discipline, the child gets possession of facts and principles—of elementary knowledge, it may be said—which are essential in his progress towards scientific concepts and organized knowledge. It seems absurd, or worse than absurd, to insist on thoroughness, on perfect number concepts, at a time when perfection is impossible, and to ignore the conditions under which alone perfect concepts, can arise—the wise working with imperfect ideas till in good time, under the law connecting idea and action, facile doing may result in perfect knowing. Following the nonpsychological method hinders the natural action of the mind, and fails to prepare the child for subsequent and higher work in arithmetic. The rational method, promoting the natural action of the mind by constructive processes which use number, leads surely and economically to clear and definite ideas of number, and thoroughly prepares for real and rapid progress in the higher work.

The Starting Point.—It is commonly assumed that the child is familiar with a few of the smaller numbers—with at least the number three. He has undoubtedly acquired some vague ideas of number, because he has been acting under the number instinct; he has been counting and measuring. But he does not, because he can not, *know* the number. He knows 3 things, and 5 or more things, when he sees them; he knows that 5 apples are more than 3 apples, and 3 apples less than 5 apples. But he does not know three in the mathematical or psychological sense as denoting measurement of quantity—the repetition of a unit of measure to equal or make up a magnitude—the ratio of the magnitude to the unit of measure. If he does know the number 3, in the strict sense, it is positively cruel to keep him drilling for months and months upon the number five and "all that can be done with it," and years upon the number twenty.

The Number Two.—There can be little doubt that among the early and imperfect ideas of number the idea of two is first to appear. From the first vague feeling of a *this* and a *that* through all stages of growth to the complete mathematical idea of two, his sense experiences are rich in twos: two eyes, two ears, two hands, this side and that, up and down, right and left, etc. The whole structure of things, so to speak, seems to abound in twos. But it is not to be supposed that this common experience has given him the number two as expressing order or relation of measuring units. The two things which he knows are qualitative ones, not units. Two is not recognised as expressing the same relation, however the units may vary in quality or magnitude; it is not yet one + one, or one taken two times—*two* apples, *two* 5-apples, *two* 10-apples, *two* 100-apples; or *two* 1-inch, *two* 5-inch, *two* 10-inch, *two* 100-inch units; in short, *one* unit of measure of any quantity and any value taken *two* times. But his large experience with pairs of things, and the imperfect idea of two that necessarily comes first, prepare him for the ready use of the idea, and the comparatively easy development of it. There must be a test of how far, or to what extent, he knows the number two. This is supplied by constructive exercises with things in which the idea of two is prominent. The child separates a lot of beans (say 8) into two equal parts, and names the number of the parts *two;* separates each part into two equal parts, and names the number of the parts *two;* separates each of these parts into two equal parts, and names the number of the single things *two.* Or, arranging in perceptive forms, how many *ones* in ⋮ ? How many *twos* in ⋮ ? How many pairs of *twos* in ⋮ | ⋮ ? Similar exercises and questions may be given with splints formed into two squares, and into two groups of two pickets each; with 12 splints formed into two squares with diagonals (see page 106); then each square (group) formed into two triangles; how many squares? how many triangles (unit groups)? how many *pairs* of triangles? Exact measurements are to accompany such exercises: the 12-inch length measured by the 6-inch, this again by the 3-inch; how many 6-inch units in the whole? 3-inch units in the 6-inch units? *pairs* of 3-inch units in the whole? Put 1-inch units together to make the 2-inch unit, the 2 inch units to make the 4-inch unit, two of these to make 8 inches, etc. It is not meant that these exercises are to be continued till the number two is thoroughly mastered; they carry with them notions of higher numbers, without which a conception of two can not be reached. Beware of "thoroughness" at a stage when thoroughness (in the sense of complete mastery) is impossible.

The Number Three.—The number three is a much more difficult idea for the child. As in the case of two, he knows three objects as more than two and less than four; three units in exact measurement as more than two, etc. But he does not know three in the strictly numerical sense. He may know two fairly well—as a working notion—without having a clear idea of the ordered or related ones making a whole. In three, the ordering or relating idea must be consciously present. It is not enough to see the three discriminated ones; they must at the same time be *related*, unified—a first one, a second one, a third one, three *ones*, a one of three. Three must be three *units*—measuring parts of a qualitative whole—units made up, it may be, of 2, or 3, or 4 . . . or *n* minor units. If 12 objects are counted off in unit-groups of 4 each, 15 objects into unit-groups of 5 each, 18 objects into unit-groups of 6 each, 30 objects into unit-groups of 10 each, the number in each and every case must be recognised as *three*. The number three, in fact, may be taken as the test of progress towards the true idea of number, and of the child's ability to proceed rapidly to higher numbers and numerical relations. If he knows three, if he has even an intelligent working conception of three, he can proceed in a few lessons to the number ten, and will thus have all higher numbers within comparatively easy reach.

The child is not to be kept drilling on the number three until it is fully mastered. What was said of two applies equally to three: it can not be mastered without the implicit *use* of higher numbers. But, while dealing with higher numbers, three may be kept in view as the crucial point in the development of exact ideas of number. There are to be constructive exercises with various measuring units in which the threes are prominent. In getting the number two, there has been practically a use of three; for where two is pretty clearly in mind, four is not far behind it in definiteness. When, for example, it is clearly seen that *two* 2-inch units make up the 4-inch unit, and two 4-inch units make up 8 inches, the two 2-inch units are perceived as *four*—i. e., ⁞ and ⁞ are seen as ⁞ | ⁞ . But for the complete conception of *four* it must be *related* not only to two (⁞), but also to three ⁞⁞ ; in other words, there must be rational counting—we *must pass through the number three* to complete recognition of the number four.

But, as already suggested, there are to be special exercises in *threes*. The dozen splints are to be used in constructing squares and triangles. How many squares? How many splints needed to make one triangle? Each of the two 6-splints is to be made into as many triangles as possible. How many triangles in each group? How many in all? "*Two* twos, or three triangles and one more." Each of the two 6-splints is to be made into as many pickets (∧) as possible. How many pickets in each group? How many in all? "*Two* 3-pickets." In the two 3-pickets how many 2-pickets? Thus, also, with exact measurements. The 4-inch units in the foot, the 3-inch units in the foot and in the 9-inch measure, the 2-inch units in the 6-inch measure; the number of 3 square-inch units in the 3-inch square, the number of 4-inch units in the 3-inch by 4-inch rectangle, the number of 3-centimetre units in the 9 centimetres, etc.

Other Numbers to Ten.—Just as when the child has a good idea of two he implicitly knows four, so when he has a good idea of three he has a fair idea of six as two threes. In ∵∵ (symbolizing any units whatever) the two 3-units are *perceived* as six units. This perception is connected with 1, 2, 3, 4, of which the child has already good working ideas, and has only to be *related* with the number five in order to fix its true place in the sequence of related acts, one, two . . . six, which completes the measurement. In this we pass through five; five is the connecting link between four and six in the completed sequence. Attention to the perception of *six* units in ∵∵ discriminates the five units ∵· from both the four and the six, and conceives their proper relation to the four units which are a part of them, and to the six units of which they are a part. When there is a fair idea of four it is an easy step to a fair idea of eight; $4 + 4 = 8$. ⁞ is not much more difficult than $2 + 2 = 4$—⁞. From the examination of eight comes the perception of seven, and a conception of its relation as one more than seven and one less than eight. From five, ⁞, it is by no means a difficult step to ten, as two fives, ⁞ | ⁞; and a comparison of this with eight readily leads to the perception of nine, ⁞ | ⁞, and to a conception of its relation to both eight and ten.

Is Ten twice Five?—From the error of considering the unit as a fixed thing, and number as arising from aggregating one by one other isolated things, arises apparently the fallacious idea that to master ten, for instance, is twice as difficult a task as to master five. Hence the prevailing practice of devoting six months of precious school life in wearying and repulsive "analyses of the number five," and finding "a year all too short" for similar analyses of the number ten. But it must be clear that by proper application of the measuring idea, wisely directed exercises in parting and wholing which promote the original activity of the mind in dealing with quantity, it is easy and pleasant to get elementary conceptions of number in general in the time now given to barren grinds on the number ten; not scientific conceptions, indeed, but sufficiently clear working conceptions, capable of large and free applications in the measurement of quantity—applications which alone can make the vague definite, and at last evolve a perfect conception of number from its first and necessarily crude beginnings in sense-perception.

The Sequence in Primary Number Teaching.—
The measuring exercises suggested, operations of separating a whole into equal parts and remaking the whole from the parts, correspond to the processes of multiplication and division. But they are not these processes, though they are presentations of sense facts which help the unconscious growth towards the conscious recognition of these processes as phases in the development of the measuring idea. It has been shown that addition and subtraction, multiplication and division, differ in psychological complexity; that each operation in the order named makes a severer demand on attention; and that, therefore, while operations corresponding to the higher processes may—indeed, must—with discretion be used from the first, they should not be made the object of conscious or analytic attention. We separate a quantity into parts—the presentative element in division; we put the parts together again—the presentative element in multiplication. But the ideas of subtraction are involved in these operations—because division and multiplication are involved—and as demanding less power of discrimination and relation are the first to be analytically taught. The separation into parts is not enough; the mere putting together of the parts is not enough. There must be a *counting* of the parts making up the whole: one, two, three . . .; and this means addition by ones. We may separate any 12-unit quantity into equal 2-unit parts; we do not know that there are *six* parts till we have counted the parts, one,

two . . . six—that is, a *first* one, a *second* one . . ., *six* ones.

Nor will the most ingenious presentations of sense material free us from this fundamental process. Such presentations help to make the process rational; they do not supersede it. If a good working idea of four has been got, it carries with it the idea of three. But it does not follow that seven (4 + 3) is known without counting. The presentation ∷ includes the presentation ∴; but the mere perception of ∷ | ∴ does not give the number seven. We perceive the four and the three, and we *know* these numbers because we have previously analyzed each of them, and put its units in *ordered relation* to one another—that is, counted them. We have a perception of the group which represents the union of these two numbers, but we do not really know *seven* till we put the three additional units in ordered relation to the four; or, in other words, till (starting with four) we count five, six, seven, thus fixing the places of the new units in the sequence of acts by which the whole is measured. We are not to rest satisfied, on the one hand, with mechanical counting—mere naming of numbers—nor, on the other hand, with mere perceptions of units unrelated by counting. This consciously relating process gives the *ordinal* element in number; counting, for example, the units in a seven-unit quantity, when we reach four we must recognize that four is the *fourth* in the sequence of acts by which the whole is constructed.

The following points in the development of number seem, therefore, perfectly clear:

ON PRIMARY NUMBER TEACHING. 181

1. The measuring idea should be made prominent by constructive exercises such as have been suggested.

2. There must be rational counting—relating—of the units of measurement. This is addition by ones. It is impossible to know, for example, ten *times*, without having added ten ones.

3. While use may—rather should—be made of the ratio idea (division and multiplication), the mastery of the combinations to ten should be kept chiefly in view; that is, addition and subtraction, with the emphasis on addition, should be first in attention, but with exercises in the higher processes. This *must* be the course if the ideas of unit and number are to be rationally evolved. In counting up, for example, the four 3-inch units in the foot measure, the child first feels and at last sees that *one* unit is 1 out of four; *two*, 2 out of four; *three*, 3 out of four; or 1-fourth, 2-fourths, 3-fourths of the whole.

4. To help in this relating as well as in the discriminating process, rhythmic arrangements of the actual units, and of points or other symbols of them, may be used with remarkable effect. The real meaning of five, as denoting related units of measure, will clearly and quickly be seen when the units (or their symbols) are arranged in the form ⁙ ; and so with other numbers. The *results* of the entire mental operation of analysis-synthesis, by which the vague whole has been made definite, are given in these perceptive forms.

5. Hence, during the first year at school we need not confine our instruction to 5, or 50, or 500, if we follow rational methods. The first thing, then, is to

see that the child gains not a thorough mastery but a good working idea of the number ten.* After exercises in parting and wholing, some notions of unit and number have been gained, and a formal start is made for *ten*, the instrument of instruments in the development of number. Many a 12-unit quantity has been divided into equal parts. Working ideas of the numbers from one to six have been obtained; the child works with them to make the numbers one to six more definite. If the preliminary constructive operations have been wisely directed the task is now an easy one. Not long-continued and monotonous drill on all that can be done upon the number ten is needed, but a little systematic work, with the ideas which from free and spontaneous use are ready to flash, as it were, into conscious recognition. Using the number forms for six, ⦂⦂, it will need but few exercises to make perfectly clear the real meaning of six, and then the remaining numbers 7 . . . 10 are, as means for further progress, within easy reach. In using these number percepts the "picturing" power should be cultivated. Five, six, seven as five and two, eight as two

---

* For the constructive exercises referred to, various objects and measured things may be used as counters; but since exact ideas of numbers can arise only from exact measurements, and since ten is the base of our system of notation, the metric system can be used with great advantage. The cubic centimetre (block of wood) may be taken as a primary unit of measure; a rectangular prism (a decimetre in length), equal to ten of these units, will be the 10-unit, and ten of these the 100-unit. The units may be of different colours, and the units of the decimetre alternately black and white. A foot-rule, with one edge graduated according to the English scale, and the other according to the metric scale, is a most useful help.

fours, ten as two fives, and five twos, etc., should be instantly recognised. In exercises upon the combinations of six we have the whole and all the related parts distinctly imaged. The analysis of the visual forms has been made $5 + 1 = 6$, $4 + 2 = 6$, etc. Now cover the 5; how many are hidden? how many seen? Cover the four; how many are hidden? how many seen? And so with all the combinations, taking care that the related pairs of number are seen as related—for example, $6 = 5 + 1 = 1 + 5$. This insures a repetition of the number activity in each case, and the ability to recognise, on the instant, any number; to see not only a whole made up of parts, but also the definite *number* of parts in the whole.

Combinations of the 10-Units.—It is hardly necessary to say that, with a fair degree of expertness (not a perfect mastery) in handling these twenty-five primary combinations, rapid progress may be made in the use of the higher numbers. As soon as a good idea of ten is gained, the pupil handles the tens, using as the ten-unit the decimetre already described (or other convenient measures), and going on to 10-tens, even more easily and pleasantly than he proceeded to 10 "ones." Is there any known law, save the law of an utterly irrational method, that confines the child for six months to the number five, for twelve months to the number ten, for another year to the number twenty, actually exacting two whole years of the child's school life—to say nothing of the kindergarten period—before letting him attempt anything with the mysterious thirty? If a child has really learned that five and three are eight, does he not know that 5 inches and 3 inches are 8 inches,

5 feet and 3 feet are 8 feet, 5 yards and 3 yards are 8 yards, 5 miles and 3 miles are 8 miles? Do five and three cease to be eight, or ten repudiate 5 times two when the unit of measure is changed? As a matter of fact, when the child, under rational instruction, gets a good grip of three he quickly seizes ten, the master key to number. As soon as he comes to ten let him formally practise with the tens the combinations he has learned with *ones:* the ten is a *unit*, because it is to be repeated a number of times to make up another quantity, just as much as " one " is a unit because it is used a number of times to make up a quantity. When it is *known* that five units and three units are eight units, it is known for all units of measure whatever. There is absolutely no limitation save this, that the child should have a reasonably good working idea of the *unit of measure*. If he knows, for instance, that 5 feet and 3 feet are 8 feet, he not only knows that 5 miles and 3 miles are 8 miles, but also has a good idea of the distance 8 miles, provided he has a good idea of the unit mile. So, when he has a working idea of a 10-unit quantity as compared with the 1-unit quantity which measures it, he passes with the greatest ease to a good idea of a magnitude measured by ten such 10-unit quantities. In using the metric units, previously referred to, he has analyzed the decimetre prism, has compared it with the minor unit—the cubic centimetre—has found that it takes ten of these minor units, or one of them taken ten times, to equal it, etc. He goes through the same constructive process with ten of these 10-unit measures; he puts ten of them together to make a square centimetre; he analyzes and compares; he uses

this new unit measure just as he used the minor-unit measure, the single cube; he finds that 5 units and 3 units are 8 units, 6 units and 2 units are 8 units, . . . 7 units and 3 units are 10 units, 8 units and 2 units are 10 units; and all the rest of it. He finds also, just as certainly as he found in operating with the 10 cubes, that to make up the whole he is measuring, one 10-unit has to be repeated ten times, 2 ten-units five times, and 5 ten-units two times. In fine, he *knows* as much about the whole, the ten 10-units, the 100 minor units, as he knows about the 1 decimetre, the 10 minor units; *knows* it just as well, is just as clear as to the relations of the several parts; for he has analyzed the undefined—the *unknown*—down to its known constituents, and built it up again by known relating processes. He does not *perceive* it just so well, does not image it as a quantity of a hundred parts; but, nevertheless, he has a fairly definite conception of the quantity as a measured whole.

Naming the Numbers.—In counting the tens the names of the new numbers may be given; or, rather, a name or two may be given, and the child will discover the others for himself: 1 ten, 2 tens, 3 tens, etc.; for *two* tens we have the somewhat special name twen-*ty* (twain-ten), and for *three* tens thir-*ty;* what name for *four* tens? for *five* tens? Let the pupils experience as often as possible the joy of discovering something for themselves.

While the work with tens is going on, practice may be had in the analysis of two tens, so as to lead to counting and naming the numbers from ten to twenty, twenty to thirty, etc. Here, again, the children will do something for themselves. The counters will be arranged, as before, so as to facilitate the perception of the corresponding numbers; 12 will be recognised as two 6's, as three 4's, as 2 more than 10, and so on. Eleven (1 ten and 1) and twelve (1 ten and 2) have special names. One ten and 3 is named thir*teen* (= three-teen). What is the name for one ten and four? One ten and five? One ten and six? The combinations, the part relations, will be learned through intuitions supplied by the measured counters, just as the combinations of ten were learned. The first learned group of combinations—those of ten—are used for the easy mastery of the second group, those from 11 to 19 inclusive. In ten there are: $\frac{1}{1}; \frac{1}{2}; \frac{1}{3}; \frac{2}{2}; \frac{1}{4}; \frac{2}{3}; \frac{1}{5}; \frac{2}{4}; \frac{3}{3}; \frac{1}{6}; \frac{2}{5}; \frac{3}{4}; \frac{1}{7}; \frac{2}{6}; \frac{3}{5};$ $\frac{4}{4}; \frac{1}{8}; \frac{2}{7}; \frac{3}{6}; \frac{4}{5}; \frac{1}{9}; \frac{2}{8}; \frac{3}{7}; \frac{4}{6}; \frac{5}{5}.$ These twenty-five combinations, gained through intuitions of things in the way already described, and applied also in constructive processes with the *ten-units*, render the mastery of the remaining *twenty-five* combinations (11–20) a comparatively easy task. They are: $\frac{2}{9}, \frac{3}{8}, \frac{4}{7}; \frac{5}{6}; \frac{3}{9}, \frac{4}{8}, \frac{5}{7}, \frac{6}{6};$ $\frac{4}{9}, \frac{5}{8}, \frac{6}{7}; \frac{5}{9}, \frac{6}{8}, \frac{7}{7}; \frac{6}{9}, \frac{7}{8}; \frac{7}{9}, \frac{8}{8}; \frac{8}{9}.$ These fifty combinations are the foundation of all arithmetical operations, and the intelligent mastery of them should from the first be the teacher's guide in all the psychological constructive processes already described or referred to.

The Notation of the Numbers.—The figure should follow the word as the word the idea of the number. When a child is able to tell the number of the units in any measured whole, he can use the figure that denotes the number. The early use of "figures" is not the cause of the baldly mechanical "method of symbols"; the figures are not responsible for the machinelike movements, any more than "words" are responsible for the machinelike movements in mere rote learning. In both cases the method is at fault. Words are taught not as *words*, but as mere sounds signifying nothing; so "figures" have been taught as empty signs, with which certain mystic operations may be performed. As in the one case the true method is not words without things nor things without words, but words *with* things; so in the other, it is not *number* without figures or figures without number, but number with figures. The intellectual peril comes in both cases from a method that retards and warps the spontaneous action of the mind. In getting the notation of ten and the higher numbers, it is almost a misuse of language to say that there is real difficulty. Elaborate analyses of the "method of teaching the figure 6," and the countless inane devices for teaching the notation of the numbers ten to twenty, and twenty to one hundred, should have no place, and have no place, in rational method. The figures from zero to 9 have to be taught, given authoritatively in connection with the ideas they denote. The symbol for ten is similarly given when ten is reached, and when the handling of the 10-units begins. Show the 10-unit alone—that is, *one* ten and *no* units, and state that it is expressed thus, 10. Then express *two* tens and no units; three tens and no units. The children express the entire series up to a hundred with the greatest ease. A like course may be followed for the numbers 11 to 19. This symbol (10) denotes one ten and *no* units. What will denote one ten and *one* unit? one ten and *two* units? Exactly the same course may be followed with all the higher numbers.

The Hundred Table.—Thus, with very little help, the pupil will name and write down all the numbers from 1 to 100. He will be greatly interested in constructing a table of such numbers and noticing how they are formed. The first column on the left has to be given him as expressing the numbers he has first learned; he must be told, also, how to write one ten and no units; he will then be able to construct the entire table

0	10	20	30	40	50	60	70	80	90
1	11	21	31	41	51	61	71	81	91
2	12	22	32	42	52	62	72	82	92
3	13	23	33	43	53	63	73	83	93
4	14	24	34	44	54	64	74	84	94
5	15	25	35	45	55	65	75	85	95
6	16	26	36	46	56	66	76	86	96
7	17	27	37	47	57	67	77	87	97
8	18	28	38	48	58	68	78	88	98
9	19	29	39	49	59	69	79	89	99

for himself. According to the plan suggested, he constructs the upper horizontal row—the tens—first: One ten and no units, two tens and no units, three tens and no units, etc.; then the second column, the numbers from 10 to 20; then the third column, two tens and 1, 2, 3, . . . 9 units, etc. He thus names and expresses the numbers from 1 to 99 inclusive. He can, of course, construct with equal ease the horizontal rows: one ten and 1, two tens and 1, three tens and 1 . . .; one ten and 2, two tens and 2, three tens and 2, . . . nine tens and 2; but the emphasis should be put on the consecutive numbers, counting from 1 to 100. Handling his counters, the child in a very short time will have working notions of numbers from 1 to 100, and will be able to interpret the symbol by selecting the right number of counters, or express any given number of counters by the right symbol. In a similar way the child will construct the 200 table, the 300 table, etc.

CHAPTER X.

NOTATION, ADDITION, SUBTRACTION.

Numeration and Notation.—When the pupil is rightly drilled in the constructive processes already described, he has learned what a unit is—a quantity used to measure another quantity of the same kind—and he has acquired a fair idea of number as denoting how many units make a given quantity. If the naming and the notation of numbers have gone on, step by step, as suggested, with the development of these ideas, all numeration and notation are potentially in his possession. He has learned to count, and to express his counts in symbols. He has learned the names of the numbers from one to ten; he has learned to use the 10-unit quantity—the hundred standard units—as a new unit, counting—as with the *ones*, the units of reference —*one, two,* . . . *ten;* he has learned to use *this* 10-unit —a thousand of the standard units—as a new unit of measure, and to count by thousands, *one, two,* etc. He has learned also the names of the numbers between ten and twenty, twenty and thirty, thirty and forty, etc.; between one hundred and two hundred, two hundred and three hundred, etc.; also, the names of the numbers between one thousand and two thousand, two thousand and three thousand, etc. In all this he has done a great deal for himself. With here and there an apt suggestion, he has been able to name the numbers from ten to twenty; to name the tens (three-ty = thirty, etc.) up to *ten* tens (the child will probably call it ten-ty), when he is given a new name for the new unit of measure—*one* hundred; and so on with the other numbers that he has been using.

The Symbols.—From idea to name, and name to symbol, is the order. As he names with but little aid from suggestion, so he needs but little assistance in notation. He is given the digits and the zero, and knows their significance. He knows that 1 denotes any *one* unit of measure whatever, and that 0 denotes *no* unit, no quantity. He is told that the expression for a 10-unit quantity is 10, meaning *one* 10-unit and *no one*-unit. The unit of reference, the *one*-unit, being called simply the unit, he will pass immediately to expressions for one ten and *one* unit, one ten and *two* units, one ten and *three* units, . . . one ten and *ten* units—i. e., two tens, or twenty. He will also express without any help *two* tens and *no* units, *three* tens and *no* units, etc.; then two tens and one unit, two tens and two units, up to ten *tens* and no units, which he will write at once as 100; counting up and expressing, that is, one ten (10), two tens (20), three tens (30), . . . ten tens (100), just as he has counted up and expressed the one-units, 1, 2, 3, 4 . . . 10. But he knows that the ten tens make a new unit of measure, viz., *one* hundred; he sees the significance of the 1 here, as in 1 (one-unit) and in 10, and counts and expresses his counts in exactly the same way—one hundred (100), two hundred (200), three hundred (300), etc., to ten hundred (1000). He has learned also that

14

ten of the hundred units make a new unit of measure—the *one*-thousand unit—and he now sees the significance of the 1 in this place (the *fourth* place), and can go on counting and expressing his counts of the ten-thousand units. Proceeding thus to any desired extent, he has almost, unaided, mastered the principles of the decimal system—the use of the zero, the absolutely unchanging values of the digits as numbers, the values of the units of measure denoted by any digit according to its place in the series, the single figure denoting so many one-units (so many units of reference—yard, dollar, pound, etc.), the second figure to the left so many ten-units, the third so many hundred-units, the fourth so many thousand-units, the fifth so many ten-thousand units, etc. He will see that 7, 75, 754, *always* denote seven, seventy-five, seven hundred and fifty-four respectively, the position of the figure or figures in each case giving the *unit;* and will note that, in reading the numbers expressed by the figures, the figures are always taken in groups of three—for example, in the number 745,745,745, each 745 is read seven hundred and forty-five, the difference being in the unit only: 745 of the *million*-unit, 745 of the *thousand*-unit, and 745 of the *one*-unit, the primary unit of reference.

The Decimal Point.—Since the pupil knows, if rightly taught upon the idea of measurement, that the unit of reference—the metre, the yard, the dollar, etc.—may itself be measured off in ten parts, or a hundred parts, etc., he will be curious to learn how the parts may be expressed. Knowing that 1 denotes 1 unit, or 1 ten-unit, or 1 hundred-unit, etc., according to its position, he will be eager to learn how it may be used to denote the one-tenth unit, ten of which make up the one unit (of reference); the one-hundredth unit, one hundred of which make up the unit; the one-thousandth unit, one thousand of which make up the unit, etc. He actually sees, for example, that the *metre* is divided into ten equal parts (tenths), each of these into ten parts (hundredths), etc. How are these to be expressed? He will have but little difficulty with the problem. In the quantity expressed by 111 metres (or dollars), he knows that the 1 on the right denotes one-metre, the next 1 one 10-metre unit, the third 1 one 100-metre unit; and passing from left to right, he knows that the second 1 denotes one tenth of the first, and the third one tenth of the second. Can we place another 1 to the right of the third, to denote one tenth of a metre, then another to denote 1-hundredth of the metre, etc.? Yes, if we in some way mark off the figures representing the sub-divisions of the unit (metre) from the multiples of the metre. We might distinguish the 1's denoting metres from those denoting parts of the metre by drawing a vertical line between them—thus: 111 | 111—the figures to the left of the line denoting, respectively, 1 metre, 1 ten-metre. 1 hundred-metre; and those to the right denoting 1-tenth metre (decimetre), 1-hundredth metre (centimetre), and 1-thousandth metre; and both constituting one series governed by the same law, namely, increasing throughout from right to left by using ten as a *multiplier*, and decreasing throughout from left to right by using ten as a *divisor*—i. e., one tenth as a multiplier. But, instead of such a separating line, it is more convenient to use a dot, called the decimal point, to *mark the place of the figure* expressing the single unit—i. e., the unit of reference. Thus the number expressed by the ones previously given will be expressed by 111·111. The first figure to the left of the unit-figure whose position is thus marked, for example, in 453·453 metres, denotes *tens;* the first figure to the right, *tenths;* the second figure to the left, *hundreds;* the second figure to the right, *hundredths;* the third figure to the left, *thousands;* the third figure to the right, *thousandths*, etc.

It will be readily observed, too, that the figures to the right of the decimal point are read in groups of three, just as those to the left are. As denoting a *number*, 453 is *always* four hundred and fifty-three. In this example it is on the left side of the decimal point, 453 *metres* (primary units); on the other it is 453 *millimetres*, etc. Here, as everywhere, there must be a good deal of drill, in order that the pupil may acquire perfect facility in reading and writing numbers; this means, ability to read automatically any number and its unit of measure, and similarly to express any quantity that may be named. For example: naming each period according to its unit of measure, name the first period (group of three figures) to the left—the *units* period; the second—the thousands (thousand-unit) period; the third to the left—the millions period; the first period to the right—the thousandths period; the second to the right—the millionths period. Make the figure 7 express billionths, hundred thousands, tens, tenths, billionths; make 45 express tens, hundreds, thousandths, millionths, etc. Care is to be taken to name correctly the measuring units in the periods to the right—for example, ·00573 is five hundred and seventy-three *hun-*

dred-thousandths; ·0006734 is six thousand seven hundred and thirty-four *ten-millionths*.

ADDITION AND SUBTRACTION.

Addition.—In addition, as we have seen, we work from and within a vague whole of quantity for the purpose of making it definite. If a quantity is measured by the parts—2 feet, 3 feet, 4 feet, 5 feet—we do not arrive at the definite measurement by simply counting the *number* of the parts; we have to count the number of the common unit of measure in all the parts, and so find the whole quantity as *so many times* this common unit. In learning addition, the countings are associated with intuitions of groups of measuring units, and the results stored up for practical use. The pupil who has been properly trained does not, in the foregoing example, start with 2, count in the 3 by ones, then the 4 by ones, etc., though this counting was part of the initial stage even when aided by the best arrangements of objects, by which he at last perceives that $5 + 4 = 9$, without now counting by ones. Addition may therefore be considered as the operation of finding the quantity which, as a whole, is made up of two or more given quantities as its parts. The parts are the *addends* (quantities to be added), and the result which explicitly defines the quantity is the *sum*. It follows that in every addition, integral or fractional, all the addends and the sum must be quantities of the same kind—i. e., each and all *must have the same* measuring unit. Not only is it impossible to add 5 feet to 4 minutes; it is impossible to add 5 feet to 4 rods—i. e., to express the whole quantity by a *number* (denoting so many units of measurement)—

without first expressing the addends in the same unit of measurement.

Thorough mastery of the addition tables must be acquired, and rapidity and accuracy in both mental and written work. Exercises on combinations, not with the single units only, but with the 10-units, the 100-units, the 1000-units—any units of which the pupil has acquired a fair working idea—will greatly aid (are a necessity) in attaining this knowledge of sums and differences as well as skill in its application. Practical facility in handling numbers *must* be acquired, at first with partial meaning, afterwards with full meaning of the operations. Some additional points may be noticed:

1. Use must be made of the knowledge of the tens as acquired in the way referred to; for example, if $8 + 8 = 16$, then $18 + 8 = 26$, $28 + 8 = 36$, etc., should follow instantly as a logical consequence. The pupil may *at first* "make up" to the next ten by separating, for example, 8 into $2 + 6$, giving, that is, $18 + 2 + 6 = 20 + 6 = 26$, just as at first he may get the sum $8 + 8$ through the steps $8 + 2 + 6 = 10 + 6 = 16$. But in all cases the *intermediate* step should be dispensed with as soon as possible, and the perception of the addends—for example, $28 + 8$—should instantly suggest the sum 36, no matter what the kind or magnitude of the unit that may be used.

2. In this connection it may be noticed that expertness in two-column addition, summing such numbers as 75, 68, no matter, again, what the unit may be, can be easily acquired—both acquired and used with the greatest interest. There is hardly a more interesting exercise in that "mental" practice which is *essential*

from the beginning to the end of the entire course in arithmetic, if knowledge, power, and skill are to be really and thoroughly gained. Thus, in finding the sum of 78 and 89, the mental movement would be the sum of the tens, the sum of the units, the tens and the units in the latter, the total in tens and units. Very soon the two "sums" are obtained simultaneously, and, with a little practice, the total (15 tens, 1 ten, 7 units) is named on the instant. With a degree of facility in adding by single columns, it is not far to equal facility in adding by double columns.

3. There should be also plenty of mental practice in addition (and subtraction) by equal increments. Count by 2's from 2 to 24; by 2's from 1 to 31; by 3's from 3 to 36, from 1 to 37, etc.

4. It affords excellent practice in written work to *set down separately the sum of each column*, the right-hand figure of each column-sum being placed under the column from which it is derived, and the other figures in their order diagonally downward to the left. These partial sums are then added together to obtain the total; thus:

In this example the sum of the first column is 42; the 2 is placed under the first column, and the 4 under the second column in the line

$9874	28
8768	29
3425	14
8267	23
2482	16
9341	17
2345	14
8273	20
2834	17
6443	17
7512	45
5454	15
$62052	195

$42 + 51 + 45 + 57 = 195$

below that of the 2. The sum of the second column is 51; the 1 is placed under the second column on the left of the 2, and the 5 is placed on the left of the 4. The sum of the third column is 45; the 5 is placed under the third column just to the left of the 1, and the 4 diagonally below to the left of 5. The sum of the fourth column is 57; the 7 is placed under the fourth column from which it was obtained, and the 5 next to the 4 in the line below. These partial sums are now added to get the total, $62052. Some advantages of this method may be noted :

(1) It helps the pupil to a clearer idea of the carrying process.

(2) In case of a mistake in the additions, it enables the pupil to detect the error more easily. What pupil has not felt the drudgery of having to go over the whole work in order to find where an error had crept in? By this arrangement the addition of any column can be tested independently of the addition of the preceding column, no knowledge of the "carried" number being required. If it is known, for example, that an error has occurred in the addition of the *thousands*, the error can be discovered and corrected without adding the *hundreds* to ascertain the number carried.

(3) Because the columns are added independently the result *may* be tested by adding the digits in each row, then adding these sums and comparing the total with the total obtained from adding the column-sums treated as separate numbers. These two totals ought to be equal. In the example, the sum of the digits in the first row is 28, in the second 29, etc., and the total

of these sums is 195, which is the same as the total of the column-sums, 42, 51, 45, 57.

(4) This method is especially useful in additions of tabulated numbers which are to be added both vertically and horizontally.

5. Another excellent practice, for the more advanced student, is in the addition of two numbers, beginning on the left. When the common plan of adding two numbers by beginning with the right-hand digits is becoming monotonous, the new method may be practised with an awakened interest because of its novelty, and at the same time a broader view of the arithmetical operation is obtained. The only point to be attended to is whether the sum of any pair of digits we are working with has to be increased by a *one* from some lower rank. In adding a pair of digits of any order, the student at the same time glances at the lower orders to see if a *one* is coming up from below to be added. In adding $\frac{628}{359}$, while adding 6 and 3, we see at a glance that their sum is not to be increased, and write down 9 at once; in adding the next pair, 5 and 2, we instantly see that their sum is to be increased by *one* from the sum of the next pair (9 + 8), and we instantly write down 8. The student should practise this method till he can use it with ease. He may exercise himself to any extent by writing down two numbers and finding their sum, then adding this sum to the last of the two numbers, then this sum to the preceding sum, etc. As addition is a further development of the fundamental process of counting, and is itself "the master light of all our seeing" in numerical operations, perfect facility

should be acquired, though not, as before said, by excluding all other ideas and operations till this perfection is attained. Get complete possession of addition, *with* full knowledge of *numbers*, if possible; *without* it, if necessary.

Subtraction.—Addition and subtraction are inverse operations. The one implies the other, and in primary operations the two should go together, with the emphasis on addition. Subtraction in actual operations with objects would seem logically to precede addition. If we wish to get a definite idea of a 14-unit quantity, and separate it into two known parts of 8 units and 6 units each, it seems that logically the 6 unit-quantity is taken away from the whole, and both the minor quantities are recognised as parts of the whole before the final process of constructing the whole from the parts is completed. There is no need, therefore, of making a complete separation between these two operations. On the contrary, they should be taught as correlative operations, with addition slightly prominent first for reasons already set forth.

From what has been shown as to the logical and psychological relation between addition and subtraction, it appears that subtraction is the operation of finding the part of a given quantity which remains after a given part of the quantity has been taken away. As in addition, so in subtraction, all the quantities with which we are working—minuend, subtrahend, remainder—must have the same unit of measurement. Further, as in addition we are working from and within a vague whole by means of its given parts, so in subtraction we are working from a defined whole, through a defined part, in order to make the vaguely conceived "remainder" perfectly definite.

Remainder or Difference.—From the nature of subtraction as related to addition, there seems to be no strong reason for the "important distinction" that should be noted between "taking" one number out of another and finding the difference between two numbers. We can not take away a given portion of a given quantity (to find the remainder) without conceiving this given portion as part of the whole; we can not get a definite idea of the "difference" between two measured quantities without conceiving the less as a part of the greater. If $5 is given as a part that has been taken from $9, we primarily count from 5 to 9 to find the *remainder*. If $5 and $9 are given as two quantities, we have to count from 5 to 9 to determine the *difference*. We have to conceive the $5 as a part of the $9.

If the preliminary work of parting and wholing to develop good ideas of number and numerical processes has been rationally done, there will be but little difficulty in the actual operation in formal subtraction. The following points with respect to the long-time mystic operations of "borrowing and carrying" may be noticed:

1. The operation involved in, e. g., 75 − 38, may and should be made perfectly clear by counters. The ten-unit in its relation to the unit has been made clear through many constructive acts. The mental process here, then, is indicated simply by

$$\begin{array}{r} 60 + (10 + 5) \\ -\ 30 -\ \ \ \ 8 \\ \hline = 30 +\ \ \ 2 + 5 = 37. \end{array}$$

If the pupil has acquired facility in the addition combinations, the operation of adding 10 and 5 and taking 8 from the sum (getting 7) is probably as easy—may become as easily automatic—as taking 8 from the 10 and adding 5 to the difference (getting 7). But the meaning and identity of both processes can be made perfectly clear. The pupil may find it at first a little easier to take 8 from the "borrowed" 10 and add 5 to the remainder (2), than to add 5 to the borrowed 10 and take 8 from the sum 15. But, in any case, these analytic acts are to lead to the clear comprehension of the process, and especially to its automatic use. There should be, of course, large practice in finding the differences of pairs of tens, as well as in finding their sums.

2. The second method of explaining the "borrowing and carrying" in subtraction—that of adding equal quantities to minuend and subtrahend—may be made equally clear. That the difference between two quantities remains the same when each has received equal increments, the pupil will discover for himself by "doing" such operations. In $75 - 38$ we add one ten-unit—i. e., ten ones—to the 5 ones, and subtract 8, as in the first case considered; i. e., $15 - 8$, or $10 - 8 + 5$; we then increase by 1-ten the 3 tens in the subtrahend, getting 4 tens, which we take from the 7 tens. This process is not a direct solution of the problem, but it is one that can be made quite intelligible. There appears to be but little difference in psychological complexity between the two methods. In both methods 8 is to be taken from 15—i. e., we have $10 + 5 - 8$. In the method of borrowing from the tens, we have to bear in mind, when we come to the subtraction of the tens, that the actual number

of tens to be dealt with is *one less* than the number of written tens. In the case of equal additions, we have to bear in mind that the actual number of tens to be dealt with is *one more* than the number of written tens.

3. Probably the best way to treat subtraction is the method based on the fact that the sum of the remainder and subtrahend is equal to the minuend. If we wish, for example, to find the difference between $15 and $8, we make up the 8 to 15, i. e., count from 8 up to 15, noting the new count of 7, which is the "difference" between 8 and 15. To find the difference between 45 and 38 is to find what number *added* to 38 will

make 45 : $\begin{array}{r} 45 \\ \underline{38} \\ \overline{7} \end{array}$ The 8 units of the subtrahend can not

be made up to the 5 units of the minuend ; we make it up, therefore, to 15 by adding 7 units, and put down 7 as a supposed part of the remainder. As this addition of 7 to 8 makes 15, we have 1 ten to carry to the 3 tens of the subtrahend, making it 4 tens, which requires *no* tens to make it up to the 4 tens of the minuend ; the remainder is therefore 7. Proceed similarly with 75 — 38, etc.

Take an example with larger numbers. From 873478 take 564693—that is, find what number added to the latter will give a result equal to the former. Write the subtrahend under the minuend, as in the margin, so that the figures of the same decimal order shall be in the same $\begin{array}{r} 873478 \\ 564693 \\ \overline{308785} \end{array}$

column. To 3, the right-hand figure of the subtrahend, 5 must be added to make up 8, the right-hand figure of the minuend ; this is the right-hand figure of the remainder. We add 8 to 9, making the 9 up to 17 (ten-unit), and putting down 8 as the second figure of the remainder. We carry the 1 (hundred) from the 17 (ten) to the 6 hundred in the subtrahend, making it 7 (hundred); this 7 (hundred) is made up to 14 (hundred) by adding 7 (hundred), which is set down in the third place of the remainder; carrying 1 from the 14 to the 4 (thousand) in the subtrahend, we have 5 (thousand), which is made up to 13 (thousand) in the minuend by adding 8 (thousand), and 8 is set down in the thousands' place in the remainder. Similarly, carrying 1 from the made-up 13 to the next figure, 6, of the minuend, we have 7, which requires *nothing* to make it up to 7, and a zero is therefore set down in the 10-thousands' place in the remainder; finally, 5 requires 3 to make it up to 8, and so 3 is set down as the last figure of the remainder. Using italics to denote the numbers to be set down in figures as the *remainder*, the statement of the mental process will be: 3 and *five*, eight; 9 and *eight*, seventeen; 7 and *seven*, fourteen; 5 and *eight*, thirteen; 7 and *naught*, seven; 5 and *three*, eight. After some practice the minuend-sums need not be pronounced, and we shall have simply 3 and *five*, 9 and *eight*, etc.

This method is usually adopted in making change, and may be used with great facility in making calculations involving both additions and subtractions. Thus, suppose a merchant, having $19128 in bank, cheques out the sums $2714, $996, $3952, $166, $7516, how much has he remaining in bank? The several subtrahends are arranged in columns under the minuend, just as in addition. Add the subtrahends and

$19128
2714
996
3952
166
7516
$3784

make up to the minuend in the way described, setting down the making-up number. The process is—

1st column: 12, 14, 20, 24 and *four*, 28—carry 2;
2d   "   3, 9, 14, 23, 24 and *eight*, 32—carry 3;
3d   "   8, 9, 18, 27, 34 and *seven*, 41—carry 4;
4th   "   11, 14, 16 and *three*, 19;

this makes up the 19 (thousand) of the minuend, and the whole "making-up" number, or remainder, is $3784, the amount of money the merchant has left in bank. The principle of "carrying" is exactly that of addition. We are making up, by successive partial addends, a smaller number to a greater. When we have come to 24 (tens)—for instance, in the second column in the example—we add 8 (tens) to make it up to 32 (tens), and so have 1 ten more—i. e., *three* in all—to carry to the next "making-up" column.

There seems to be no good ground for the assertion sometimes made that this method is illogical, and wastes a year or more of the pupil's time. The first statement is refuted by the psychology of number; the second, by actual experience in the schoolroom. If to think from 15 down to 7 is logical, it would be no easy task to show that to think from 8 up to 15 is illogical. We can neither think down in the one case nor up in the other without thinking of a measured whole of 15 units as made up of two parts, one of 7 units, the other of 8 units. As a *conscious process*, $8 + 7 = 15$ carries with it the inevitable correlates $15 - 8 = 7$, $15 - 7 = 8$. From what has been shown as to the relations of the fundamental operations, it might even be inferred that if there is any difference in difficulty between the making-up method and the taking-away method, the difference is in favour of the making-up method, as involving less demand upon conscious attention. However this may be, it is certainly known from actual knowledge of school practice that pupils who have been instructed under psychological methods have had but little difficulty in comprehending the making-up method, and have quickly acquired skill in the application of it.

Fundamental Principles of Addition and Subtraction.—When a quantity is expressed by means of several terms connected by the signs + and −, the expression is called an *aggregate;* and when the several operations are performed the result is the *total* or sum of the aggregate. Some of the fundamental principles connecting the operations of addition and subtraction are :

(1) If equals be added to equals, the wholes are equal.

(2) If equals be subtracted from equals, the remainders are equal.

(3) Adding or subtracting zero from any quantity leaves the quantity unchanged.

(4) *Changing the order* of performing the additions and subtractions in any aggregate *does not change* the total or sum of the aggregate.

The pupil can use these principles, and abstract recognition of them will come in good time.

CHAPTER XI.

MULTIPLICATION AND DIVISION.

Multiplication.—From the preceding discussion (see especially page 109 *et seq.*) of multiplication as a stage in the development of number, it is clear that certain points are to be kept steadily in view, if the *process* is to be made really intelligible to the pupil.

1. It is not simply addition of a special kind. It means development and conscious use of the idea of *number*—that is, of the ratio of the measured quantity to the unit of measure, whatever the magnitude of the unit may be in terms of minor units. In counting with a 1-unit measure, one, two, three, . . . nine, the number is known when the unit it names is recognised as the *ninth* in a series of nine units constituting a whole—when, that is, the defined quantity is grasped as nine times the unit of measure.

2. In the development of the measuring process (as in the exact stage of measurement) there is the *explicit recognition* that the measuring unit is itself measured off into a definite number of minor units. This gives rise to the process of multiplication, and of course to a more definite and adequate idea of *number* as denoting times of repetition of the unit to make up or equal the magnitude. Nine times *one* is *nine* is understood in its full significance.

3. A quantity expressed in terms of a given unit of measure is, by multiplication, expressed in terms of the minor units in the given unit of measure; in other words, for the number of derived units in the quantity is substituted the *number* of primary units in the quantity. If we buy 7 barrels of flour at $5 a barrel, the measured cost is $5 × 7; *seven* units of $5 each. By multiplication this is changed to $35—i. e., $1 × 35. This product, as it is called, this new measurement, is not *seven fives*. It denotes the same quantity under a different though related measurement; it is thirty-five *ones*. In one of these measurements the *number* is seven, in the other it is thirty-five.

4. The multiplicand must always be regarded as a unit of measure—a measure made up of primary units; and the operation looked upon as simply making the quantity more definite by expressing it in a better known or more convenient unit of measure.

5. While the multiplicand as multiplicand must always be interpreted to mean measured quantity, we can take either factor as multiplier or multiplicand. This idea must be used from the first, even in the primary stage. In finding the number of primary units (dollars) in 12 yards of velvet at $5 a yard, there is no known law that decrees 12 as unchangeably the multiplier, and $5 as the only multiplicand. On the contrary, by a necessary law of mind, every measuring process has two phases, and so the measurement $5 × 12 carries with it the measurement $12 × 5. Only a total misconception of number and the measuring process could prompt the question, How can 12 yards become $12? The proposition $5 × 12 = $12 × 5, is not a proposition

about *things;* it is a proposition concerning a psychical process—the mind's mode of defining and interpreting a certain quantity. This principle of measurement—of interchange of times and parts—is essential to the proper understanding of numerical operations, and can from the beginning be intelligently used. Intelligent use leads to perfect mastery. The problem of multiplication then is: Given the *number* of unit-groups in a measured quantity, and the *number* of minor units in each unit-group, to determine, from these related factors, the number of minor units in the quantity.

The Formal Process of Multiplication.—It may be well to consider the logical steps in learning the process:

(1) The multiplication of a quantity by powers of ten. Beginning with some ultimate or primary unit of measure, we conceive a measured quantity as making up ten such units—that is, we multiply the unit by ten; we may further conceive this 10 unit quantity used as a unit of measure, and repeated ten times to make up a larger quantity—that is, the 10-unit quantity is multiplied by ten to express this larger quantity in terms of the minor unit, it is 100 of them, etc. It has already been shown how the notation corresponds with this process. The 1-unit multiplied by 10 becomes 10, the 10-unit multiplied by 10 becomes 100; in other words, the 1 increases 10 times with every removal to the left of the decimal point. So the product of 5 ones is 10 fives or 5 tens—i. e., 50; the product of 5 tens by 10 is 50 tens or 500—i. e., 5 multiplied by 100, etc.

(2) We may find the total product which measures

a quantity by finding the sum of partial products. If a given quantity is measured by 4 feet × 28, we may multiply the 4 feet by 20 and by 8, and the sum of the partial products will be the total product—in all 28 times the multiplicand. This is the basis of the work in long multiplication.

(3) We may multiply by the factors of the multiplier. This is using the relation between parts and times. If we have, e. g., a quantity expressed by $2 × 20, it is expressed equally by $2 × 5 × 4—i. e., by $10 × 4; in other words, we have made the measuring unit 5 times as large, and the *number* of them 5 times as small.

In the following, e. g., we take the multiplicand 5 times, getting the first partial product; in multiplying by 4, we have in fact taken the multiplicand 10 times, and this product 4 times, obtaining the second partial product 11,120.

```
   278
    45
  ────
  1390  . . .  5 times multiplicand.
  1112  . . .  40 (i. e., 4 times 10 times multiplicand).
  ─────
  12510  =  45 times multiplicand.
```

Special Processes.—Special processes may be used in many cases. These afford good practice for mental work, and give better ideas of number and numerical operations, as well as preparation for subsequent work. A few of these processes may be noticed.

1. When the multiplier is any of the numbers 11 to 19, the product can be obtained in one line, thus:

MULTIPLICATION AND DIVISION.   211

8765	Nine 5's, 45—carry 4;
19	" 6's, 58 and *five*, 63—carry 6;
166535	" 7's, 69 and *six*, 75—carry 7;
	" 8's, 79 and *seven*, 86—carry 8;
	8 and *eight*, 16.

The number in italics is in each case the number in the multiplicand just to the right of the one multiplied. To multiply by 31, 41, 91, it is best to write the multiplier over the multiplicand, and use the multiplicand itself as the partial product from the digit 1 in the multiplier. For example:

$$\begin{array}{r} 81 \ldots \text{the multiplier.} \\ \hline 96478567 \ldots \text{product by 1.} \\ 771829536 \ldots \text{product by 8 (tens).} \\ \hline 7814773927 \ldots \text{product.} \end{array}$$

The product can be obtained in one line, as in multiplying by 19, but there is greater risk of error in the mental working. Such examples as 84×76 afford interesting and useful mental practice. Multiplying crosswise and summing the products, 76 *tens*; multiplying the units, 2 *tens* 4 units; multiplying the two tens, 56 hundreds; hence 63 hundreds, 8 tens, and 4 units—i. e., 6384.

2. Practice in finding the squares of numbers is very useful. The rule for finding the square of the sum of two numbers and the difference between the squares of two numbers may be readily arrived at. For example, multiply 85 by 85:

$$\begin{array}{r} 80 + 5 \\ 80 + 5 \\ \hline 80^2 + 5 \times 80 \\ + 5 \times 80 + 5^2 \\ \hline 80^2 + 10 \times 80 + 5^2 = 7225 \end{array}$$

This may be illustrated by intuitions, symbolising units by dots. Let the following indicate the square of 7  (5 + 2). We see at once that to make up the whole square there is (*i*) the square of 5, (*ii*) 5 taken *twice*, and (*iii*) the square of 2—that is, the square of the first number, twice the product of the first by the second, and the square of the second number. It will be readily seen that the difference of the two squares ($7^2 - 5^2$) is *twelve times two;* but twelve is the *sum* of the numbers and *two* their difference. Does this hold for other numbers? The pupil will be greatly interested in discovering for himself the general principle: the difference of the squares of two numbers is equal to the sum of the numbers multiplied by their difference.

If, in the figure, he compares the square of 3 with the square of 4, of 4 with that of 5, he will see that the square of any of these numbers is got from the square of the next lower by simply adding the *sum* of the numbers to the square of the lower. The square of 3 is 9; the square of 4 is 9 ($= 3^2$) + (3 + 4); the square of 5 is 16 + (4 + 5); and the square of 7 is 36 + (6 + 7). The pupil will deduce for himself that, given the square of any number, the square of the next consecutive number is obtained by adding the sum of the numbers to the given square.

All these principles, and many others, may be made the basis of exercises equally interesting and useful in mental arithmetic: Square of 95; multiply 95 by 105, (100 − 5) (100 + 5); 295 by 305, (300 + 5) (300 − 5); the square of 250; the square of 251, etc.

MULTIPLICATION AND DIVISION. 213

3. The making-up method in subtraction may be conveniently used when the product of one number by another has to be taken from a given quantity.

From 89713 take 8 times 8793. The work is done as follows:

89713	Eight 3's, 24 and *nine* = 33—carry 3;
8793	" 9's, 75 and *six* = 81—carry 8;
8	" 7's, 64 and *three* = 67—carry 6
19369	" 8's, 70 and *nineteen* = 89.

The numbers in italics indicate the remainder, 19369.

4. Advantage may often be taken of the fact that some of the numbers (tens, etc.) of the multiplier—and, once more, *either* factor may be made the multiplier—represent a multiple of some of the others. If, for instance, we want to find the cost of 2053 bags of flour, at $3.287 a bag, we may use the latter for multiplier, and write only three partial products:

$$\begin{array}{r}
2053 \\
3287 \\
\hline
14371 \ldots\ 7 \text{ times.} \\
57484 \ldots\ 280 \text{ times.} \\
6159 \ldots\ 3000 \text{ times.} \\
\hline
\$6748.211
\end{array}$$

In this example we multiply by 7, and, observing that 28 is 4 times 7, we multiply the first line of the product by 4, getting the second line; then the multiplicand by 3, taking care, of course, to put the product in the thousands' place.

We may often take advantage of this method by breaking the order of finding the partial products.

Thus, if the product of 567392 by 218126 is required, we may use the former as multiplier, and work thus:

$$
\begin{array}{r}
218126 \\
567392 \\
\hline
1526882 \quad \ldots 7000 \text{ times.} \\
12215056 \quad \ldots 560,000 \text{ times.} \\
85505392 \ldots 392 \text{ times.} \\
\hline
123{,}762{,}947{,}392
\end{array}
$$

We notice that 56 is 8 times 7, and that 392 is 7 times 56. Begin, therefore, with 7. Multiplying this product by 8, we have the second line of partial products; and, finally, multiplying this second line by 7, we get the third line of partial products.

Or we might have used 218126 for multiplier, observing that 9 times 2 are 18, and 7 times 18 is 126; thus:

$$
\begin{array}{r}
567392 \\
218126 \\
\hline
1134784 \\
10213056 \\
71491392 \\
\hline
123{,}762{,}947{,}392
\end{array}
$$

Multiplying first by 2 (hundred thousand); multiplying this product by 9; multiplying this second partial product by 7, taking care as to the proper placing of the products, we have the complete product.

5. Another plan that affords a good exercise in mental additions, and subsequently proves useful, is the method of finding a product of two factors in a single line. To multiply, e. g., 487 by 563, write the

MULTIPLICATION AND DIVISION. 215

multiplier, with the digits in inverted order, on the lower edge of a slip of paper, thus, |3 6 5|. Place the paper over the multiplicand so that the units (3) shall be just over the units of the multiplicand. The artifice consists in moving the slip of paper along the multiplicand, figure by figure, till the last digit (5) of the inverted multiplier is over the last digit of the multiplicand, and taking the product of any pair, or the sum of products of any pairs, of numbers that may be in column. Thus:

|3 6 5|
  4 8 7

Three 7's, 21—*one* and carry 2;

  |3 6 5|
   4 8 7

Three 8's, 26; six 7's, 68—*eight* and carry 6;

   |3 6 5|
    4 8 7

Three 4's, 18; six 8's, seven 5's; 101—*one* and carry 10;

    |3 6 5|
     4 8 7

Six 4's, 34; five 8's; 74—*four* and carry 7;

     |3 6 5|
      4 8 7

Five 4's and 7 carried—*twenty-seven*.

The numbers in italics, taken in order, are the product, 274181.

Proofs of Multiplication.—(1) By repeating the operation with the factors interchanged. (2) The product divided by either factor should give the other factor. (3) By casting the nines out of the multiplier and the multiplicand, then multiplying these remainders together and casting the nines out of their product; the remainder thus obtained should equal the remainder from casting the nines out of the product of multiplier and multiplicand. For example, test the following by casting out the nines:

$$987761 \times 56789 = 56{,}093{,}959{,}429.$$

$$\begin{aligned}
&\phantom{\text{Out of multiplicand ..}} 7 \ \ \text{.. out of product of 2 and 8.} \\
&\text{Out of multiplicand ..} 2 \times 8 \ \text{.. out of multiplier.} \\
&\phantom{\text{Out of multiplicand ..}} 7 \ \ \text{.. out of product.}
\end{aligned}$$

This proof is not a perfectly sure test of accuracy. It does not point out an error of 9, or of a multiple of 9, in the product. Thus, if 0 has been written for 9 or 9 for 0, or if a partial product has been set down in the wrong place, or if one or more noughts have been inserted or omitted in any of the products, or if two figures have been interchanged, or if one figure set down is as much too great as another is too small, casting out the nines will fail to detect the error, for the remainder from dividing by 9 will not be affected. Still the proof is interesting, as throwing light upon the decimal system of notation.

The Multiplication Table.—The sure groundwork for this table is, of course, facile mastery of the addition and subtraction tables. Though scraps of it given from time to time—as the 2's and 3's in 6—are perhaps of no great value as contributing to the making and mastering of the entire table, yet some complete parts of the table—as, for example, two times, three times, ten times—may be kept in view, and may be expertly handled quite early in the course. It has been said that the table is a grand effort of the special memory for symbols and their combinations, and that the labour can not be extenuated in any way. The labour is, indeed, heavy enough, but it is believed that it may be somewhat lightened. The table, as the key to arithmetic, must be learned, and it must be learned perfectly—i. e., so that any pair of factors instantly suggests the product; there must be no halting memory summoning attention and judgment to its aid. It is therefore worth while to "extenuate" the labour of learning it, if this can possibly be done. To this end some suggestions are made which are believed to be rational, while they have certainly stood the test of experience.

1. *The Meaning of the Table.*—Pupils rightly taught know how to construct the table; they know what it means. The symbol memory, like every other kind of memory, is always aided where intelligence is at work. In former times, not so long past, the table used to be said or sung—rattled off in some familiar tune—without a glimmer of what it all meant; but under rational instruction the children know several important things about it, and the teacher should use these things in lessening the labour of complete mastery.

2. *Memory aided by Intelligence.*—(1) The pupils have learned how to construct any part of the table, two times, three times, etc.

(2) They know exactly what such construction means, for they have acquired a fair idea of times—of number as denoting repetition of a measuring unit. They know, therefore, the meaning of every product: 6 oranges at 5 cents apiece, 6 yards of calico at 9 cents a yard, etc.

(3) They can derive the product of any pair of factors from the product of the immediately preceding pair. Knowing that 6 yards of cloth at 8 cents a yard cost 48 cents, they know that 7 yards cost 8 cents more. Similarly they quickly learn that if 10 oranges cost 50 cents, 9 oranges will cost five cents less, and 8 oranges *one* ten less, etc. Thus they will have various ways of constructing, and recovering when momentarily forgotten, the product of any pair of digits.

3. *The Commutation of Factors.*—In learning the table the relation of the factors must be kept in view. This greatly reduces the labour. There ought to be little difficulty in this if a fair idea of the relation between parts and times has been brought out. At 3 cents apiece, 5 oranges cost 3 cents × 5; this is seen to be identical with 5 cents × 3. Each of these expresses measured quantity, a sum of money; the thought "oranges" disappears from this conception. The table is often taught without reference to this principle, and so the labour of learning it is at least one half greater than it ought to be. In our boyhood we learned 9 × 6 = 54, without a suspicion that 6 × 9 = 54. Let us see to it that the present things be made better than the former.

4. *Memory further aided.—Associations.* In this connection the following suggestions are worthy of attention:

(1) The thing to be kept in view is that, so far as possible, *associations are to be formed directly between a product and one or both of the factors which produce it.*

(2) *Ten* times is already learned in addition—in the counting of the tens. The pupil knows how to "multiply" any number by ten by simply affixing a zero to the number. The association of product and factors is direct; the product is the multiplicand with the zero of the 10 affixed. Ten times, then, is well in hand.

(3) *Eleven* times is almost equally easy. The product in each case, for the first nine digits, is directly associated with the *digit;* the digit is simply repeated— 11, 22, 33, etc. Eleven tens is known from ten elevens, and the other two products (11 × 11, 12 × 11) must be built up from this.

(4) *Nine* times can be formed and remembered in a similar way. The pupil will note: (*a*) That a product is made up of tens and units. (*b*) That in 9 times (up to 10 × 9) the *number* of tens is always one less than the *number* multiplied. (*c*) That in every product the sum of the digits is 9; and thus, having written down the *tens* directly from the multiplicand, he can at once write the units. He should be led to notice also that (*b*) holds good as to the law of tens up to 10 × 9, after which the number of tens is *two* less than the multiplicand up to and including 20 × 11, after which the number of tens is *three* less, etc. He should note, too, in his formed table, how the tens increase by one and the units decrease by one. This may seem somewhat complex, but it works well. We have known a boy of six years to construct and learn 9 times up to 9 times 10 in fifteen minutes.

(5) Probably *two* times has been completely learned before a formal attack is made upon the table as a whole. There has been much practice in counting by 2's—backward and forward—and by 3's, etc. There seems to be no way of making a mnemonic association between a product and its factors; but addition by two is an easy operation, and two times is quickly learned.

(6) In *twelve* times (assuming two times) the memory can be aided by association. The product of any multiplicand may be obtained by taking 12, the multiplicand, as so many *tens*, and doubling it for the units; twelve times $3 = three$ tens and *six* (twice 3) units. For 5 and up to 9, doubling the unit gives more than 10, but the additions are easy. 12 times $5 = five$ tens and *ten* units (twice *five*) = 60. Or, consider the products

thus: the products of 1 — 4 are 12, 24, 36, 48; those of 5 — 9 are each *one* ten more than the multiplicand, and the units increase by 2's—i. e., 0, 2, 4, 6, 8; the products of 5, 6, 7, 8, 9 are therefore 60, 72, 84, 96, 108.

(7) Some assistance from association may be had in learning 5 times by observing: (*a*) that the *units* are alternately 5 and 0, 5 for the *odd* multiplicands, 0 for the even ones; (*b*) if the multiplicand is even, the *tens* are half of it; if *odd*, the tens are *half* the next lower number: $8 \times 5 = 4$ tens and 0 units; $9 \times 5 = 4$ tens and 5 units, etc. More advanced students will take pleasure in extending the multiplication table according to these laws, as well as in accounting for the laws. For example: in 9 times, why are the tens *one* less than the multiplicand up to 10×9, then *two* tens less up to 20×9? etc. In 8 times why are the tens *one* less than the multiplicand up to 5×8, *two* less from 6×8 to 10×8? etc.

Division.—Division is, we have seen, the operation of finding either of two factors when their product and the other factor are given. After what has been said in Chapter V upon the nature of division and its relation to multiplication and fractions, little further need be added, especially as most of the text-books explain clearly enough the actual arithmetical work. A few points, however, may be briefly noticed: (1) If, in the method of teaching, the idea of number as measurement has been kept steadily in view, the nature of division as the inverse of multiplication will be fully understood. (2) Knowing the relation of the factors in multiplication, the pupil will, with but little difficulty,

MULTIPLICATION AND DIVISION.   221

comprehend the operation and be able to interpret the results in every case. Practised from the first in using the idea of correlation—of *number* defining the measuring unit and *number* defining the measured whole—in both multiplication and division, he can tell on the instant which of these factors is demanded in any problem. (3) There does not seem to be any necessity for beginning formal division by the "long division" process. The pupil knows that $2 \times 5 = 10$, and that $10 \div 5 = 2$, *whatever may be the unit of measure.* He knows that ten ones divided by 5 is two ones, that ten tens divided by 5 is two tens, ten hundred-units divided by 5 is two hundred-units, etc. He has learned that 12 units of any order in the decimal system when divided by 5 gives 2 units of that order, with 2 units of that order, or 20 units of the next lower order, remaining; which 20 units on division by 5 gives 4 units of that order, making the total quotient 24. In short, if the pupil has been taught to divide a number of any two digits by any of the single digits, he can divide any number by a single digit. Thus, suppose 4976 is to be divided by 8:

8)4976
  622

here eight will not divide 4 giving a quotient of the same order—i. e., in the thousand units; the 4 is changed to 40 units of the next lower order, making, with the 9 of that order, 49. This divided by 8 gives 6, with 1 over. Similarly this 1 is 10 of the next order, which, with the 7 of that order, makes 17; this divided by 8 gives 2, with 1 over; this 1 is 10 of the next order, and with the 6 makes 16 of that order, which, divided by 8, gives 2, the last figure of the quotient. No matter what the series of figures, the process is the same, and the pupil should experience no real difficulty if rational method and practice have been followed. A few practical points may be noted:

(1) The division by any power of 10 is as easy as multiplication by any power of 10—is, in fact, derived directly from it.

(2) So with division by factors of the divisor, which is directly connected with multiplication by factors of the multiplier. To the pupil it will prove an interesting exercise to discover the "true remainder." Take, for example, $5795 \div 48$.

8)5795
6)724 ... 3 *rem.* in *ones*, the quotient being 724 *eights*.
120 ... 4 *rem.* in 8-unit groups;

hence remainder in *ones* is $8 \times 4 + 3 = 35$. This is the old rule: Multiply the first divisor by the second remainder and add the product to the first remainder. The same method is applicable to the case of three or more factorial divisors; apply the rule to the last two divisions, and use the result with the first divisor and first remainder. Or, reduce each remainder to units as it occurs; for example, divide 2231 by 90 ($= 3 \times 5 \times 6$).

3)2231
5)743 unit-groups of 3 with *rem.* 2 units;
6)148 unit-groups of 15 with *rem.* 3 groups of $3 = 9$ units;
24 groups of 90 with *rem.* 4 groups of $15 = 60$ units.

The remainder is therefore $60 + 9 + 2 = 71$. Otherwise, applying the rule with the last two divisions: $5 \times 4 + 3 = 23$; use this as the "second remainder" from the "first divisor," and remainder $23 \times 3 + 2 = 71$.

(3) In long division the multiplications and subtrac-

MULTIPLICATION AND DIVISION.   223

tions may be combined, as described under multiplication and subtraction—e. g., 635040 ÷ 864.

$$864 \overline{)635040} (735$$
$$\underline{3024}$$
$$4320$$

(1) Seven 4's, 28 and *two* = 30—carry 3. (2) Seven 6's, 45 and *zero* = 45—carry 4. (3) Seven 8's, 60 and *three* = 63. This gives 302, which, with 4 brought down, makes the first remainder. Proceed similarly with 3 and 5, the other figures in the quotient. The student may note the application of the method in a longer operation: Divide 217,449,898,579 by 56437. The following is the work:

$$3852967$$
$$56437 \overline{)217449898579}$$
$$481388$$
$$298929$$
$$167448$$
$$545745$$
$$378127$$
$$395059$$

Three 7's, 21 and *eight* = 29—carry 2. Three 3's, 11 and *three* = 14—carry 1. Three 4's, 13 and *one* = 14 —carry 1. Three 6's, 19 and *eight* = 27—carry 2. Three 5's, 17 and *four* = 21. This gives 48138, which, with the **8** (heavy-faced type) brought down, makes the complete first remainder. With this proceed exactly as before, and so on with the other remainders.

(4) *Casting out the Nines.*—It is seen that 9 (and

16

of course 3) is a measure of 9, 99, 999, 9999, etc.—that is, of $10 - 1$, $100 - 1$, $1000 - 1$, etc. Hence, if from any number there be taken all the *ones*, and 1 from *every* 10, 1 from *every* 100, etc., the remainders from the tens, the hundreds, the thousands, etc., constitute a number which is a multiple of 9. The original number will therefore be a multiple of 9, if the total of the deductions made is a multiple of 9; this total is the number of *ones* + the *number* of tens + the *number* of hundreds, etc.—that is, this total is the sum of the digits of the given number. For example, is 39273 divisible by 9?

30000 = 3	times	10000 = 3	times	9999	and	3
9000 = 9	"	1000 = 9	"	999	"	9
200 = 2	"	100 = 2	"	99	"	2
70 = 7	"	10 = 7	"	9	"	7
1 = 3	"	1 = 3	"	0	"	3

Adding 39273 = some multiple of 9 and 24

Hence the given number is exactly divisible by 3, but leaves a remainder of 6 when divided by 9, because $24 \div 9$ leaves 6 remainder. The principle is: any number divided by 9 leaves the same remainder as the sum of its digits divided by 9.

To cast the nines out of any number, therefore, is to find the remainder in dividing the number by 9. In casting out the nines from the sum of the digits we may conveniently omit the nines from the partial sums as fast as they rise above 8.

Proofs of Division.—(1) By repeating the calculation with the integral part of the quotient for divisor. (2) By multiplying the divisor by the complete quo-

MULTIPLICATION AND DIVISION. 225

tient. (3) By casting out the nines, as in multiplication. For example:

$$3{,}893{,}865{,}223 \div 179 = 21{,}753{,}437.$$

$$\begin{array}{r} 4 \quad \ldots \text{out of product } 8 \times 5. \\ \text{9's out of divisor} \ldots 8 \times 5 \ldots \text{out of quotient}. \\ 4 \quad \ldots \text{out of dividend}. \end{array}$$

If there is a remainder the method can still be applied. Test the accuracy of

$$3{,}893{,}865{,}378 \div 179 = 21{,}753{,}437 \tfrac{155}{179}$$

where the remainder is 155.

$$\text{9's out of divisor} \begin{cases} 6 \quad \ldots \text{out of } 8 \times 5 + 2. \\ \ldots 8 \times 5,\, 2 \ldots \text{out of quotient and } \textit{remainder}. \\ 6 \quad \ldots \text{out of dividend}. \end{cases}$$

The disadvantages of this proof are similar to those in the proof of multiplication by casting out the nines.

Fundamental Principles connecting Multiplication and Division.—From the theory of number as measurement and numerical operations as a development of the measuring idea, there are certain fundamental principles—fundamental also in fractions—connecting the operations of multiplication and division. The principal of these are the following:

(1) If equals be multiplied by equals, the products are equal.

(2) If equals be divided by equals, the quotients are equal.

(3) If an expression contains a series of multipliers and divisors, changing the order of the multipliers and divisors does not change the value of the expression.

The last principle includes several principles of useful application, either implied or stated explicitly in the discussions upon number and its development.

(*a*) The order of numerical factors may be changed. (*b*) Multiplying a factor by any number multiplies the product by the same number. (*c*) Dividing a factor of any number divides the product by the same number. (*d*) Multiplying the dividend by any number multiplies the quotient by that number. (*e*) Dividing the dividend by any number divides the quotient by the same number. (*f*) Multiplying the divisor by any number divides the quotient by the same number. (*g*) Dividing the divisor by any number multiplies the quotient by the same number. (*h*) Multiplying or dividing both divisor and dividend by the same number leaves the quotient unaltered. (*i*) All these principles are necessarily involved in the principles of number as already unfolded. The following is worthy of attention: In an *aggregate* whose terms contain multipliers and divisors, *the multiplications and the divisions are to be performed* BEFORE *the additions and the subtractions are made.*

CHAPTER XII.

MEASURES AND MULTIPLES.

Greatest Common Measure.—The pupil who has been led to have a clear idea of number—who has been taught to look upon the unit as the *measurer*—will find no difficulty in mastering greatest common *measure*. With all the preliminary notions he is familiar, and it will be an easy matter to pass to the formal process.

While in the illustrations given in this chapter we generally use the pure number symbols, it must be borne in mind that here, as everywhere in number and numerical processes, the idea of *measurement* is to be kept prominent, especially in the introductory lessons. A common factor is a common *measure*—a unit of measure that is contained in two or more quantities an exact number of times. A common multiple is a definitely measured quantity, which can be measured by two or more quantities, themselves measured by units of the same kind and value as those of the given quantity. The teacher should see to it, then, that all his illustrations and examples deal with the concrete; that the measuring idea be kept prominent from first to last.

Easy Resolution into Factors.—Taking the number 15, the learner sees that it can be considered 3 fives, or 5 threes; the five or the three is a measurer or measure of 15, and the equation $15 = 5 \times 3$ puts in evidence the fact that 5 and 3 are measures or factors of 15. Taking 35, he sees the significance of the equation $35 = 5 \times 7$. He further notes that 5 is a measure of each of the numbers 15 and 35, and is therefore a *common* measure. If, next, the numbers 12 and 18 are taken, he will see that all the measures of 12 are—

$$1, 2, 3, 4, 6, 12;$$

and that all the measures of 18 are—

$$1, 2, 3, 6, 9, 18.$$

Then it will be seen that 1, 2, 3, 6 are common measures of 12 and 18, and that while there are several such measures, there is one that is the *greatest*—the one that will be called the greatest common measure. Before any process is taught the class should be exercised in the working of easy examples, both mental and written; being asked to find common measures, and the greatest common measure of 16 and 24, of 24, 36, 48, etc. An additional interest will be secured by proposing some simple practical problems.

It will be better, before beginning the ordinary formal treatment, to have exercises in finding the greatest common measure, by resolving the numbers given into their simple factors. It would be necessary, then, to recall or develop a certain fundamental principle. The division $\begin{array}{r}2)\overline{60}\\3)\overline{30}\\\hline 10\end{array}$ is to be interpreted, first, that 60 is 30 twos, and, next, that the 30 twos are 10 three-twos or 10 sixes; and thus that if a number contains the factor

2, and if the quotient contains the factor 3, the number itself contains the factor or measure **6**. Then, since
$$108 = 2 \times 2 \times 3 \times 3 \times 3,$$
$$\text{and} \quad 72 = 2 \times 2 \times 2 \times 3 \times 3,$$
we may see that all the single common factors are 2, 2, 3, 3; and that, therefore, $2 \times 2 \times 3 \times 3$, or 36, is the greatest common measure. Practice on this method will find a place: the pupil has a new interest, and the teacher can take advantage of it to secure further training in number and in the elementary processes.

The General Method.—But soon it will be found that this method is limited, as its successful application depends on the pupil's ability to discover a factor. An example, such as, Find the greatest common measure of 851 and 1073, we may suppose to have been given the class, and found beyond their present power of factoring. The reason for the failure will be manifest to them—their inability to find any factor of either number. The need for some new, or, it may be, extended method, is felt; and this need is the teacher's opportunity for introducing the more powerful method, and for the development of it he has his class in a state of healthy, natural, unforced interest.

The Fundamental Principles.—To develop the method, it would be well to turn aside from the example attempted and give attention to certain facts upon which the method is based. Taking for illustration the numbers 21 and 35, we see, as before, that 21 is 3 sevens and 35 is five sevens. Thus, if 21 is added to 35 we shall have 3 sevens, and 5 sevens or 8 sevens; the seven being the unit of measure, or measurer. Similarly, if 21 is subtracted from 35 the result will be 2 sevens. Further, if to 21 is added 3 times 35, we have 3 sevens and 3 times 5 sevens—that is, a certain number of sevens. This is seen to be true for any number of times seven, any number of times eight, or nine, . . . etc. Actual measurements will make the principle still clearer. Thus, if A B and C D have a common measure, it must measure A B exactly, and C D exactly:

```
―――――    ――――――――――
A      B  C      E       D
```

and measuring off on C D a part C E = to A B, the common measure must measure C E exactly, and therefore E D exactly, because it measures the whole of C D; but E D is the difference of the quantities, etc. In the same way E D may be measured off on A B, and the same reasoning will apply. Thus the pupils are led to see certain general principles, and to see them in their generality.

1. From the fact that if we take the sum or the difference of 21 and 35—that is, of 3 sevens and 5 sevens —or the sum or the difference of any number of times 21 and any number of times 35, we are sure to have a number of sevens (seven representing any measured quantity whatever), it is plain that any number which measures two numbers will measure their sum or their difference, or the sum and also the difference of any of their multiples. The pupils can be got to develop the general form of this principle. If c is a common measure of a and b, so that $a = mc$, and $b = nc$, then $a + b = mc + nc$, etc.

2. Because the common measure of two numbers measures their sum, and because the minuend, in a sub-

traction operation, is the sum of the remainder and the subtrahend, it is plain that every common factor of the remainder and the subtrahend is a factor of the minuend.

The Application of the Method.—We pass now to the application, and shall take the numbers 851 and 1073. The difficulty has been that these numbers are large, and in reply to the question, What smaller number will have in it any common factor that 851 and 1073 may have? there might be expected the answer, 1073 − 851. But there must be an examination of this statement.

$$851 \mid \begin{array}{r} 1073 \\ 851 \\ \hline 222 \end{array}$$

If 851 and 1073 have a common factor, this factor will also measure 222; and if 222 and 851 have a common factor, this factor will measure 1073. Thus the greatest common measure of 851 and 1073 is a factor of 222, and the greatest common measure of 851 and 222 is a factor of 1073. Therefore the greatest common measure of 851 and 222 is the greatest common measure of 851 and 1073. It will now be easy to show that if 222, or 2 times 222, or 3 times 222, be taken from 851, 222 and this remainder will have for greatest common factor the greatest common factor of 851 and 222, and the advantage in taking from 851, 3 times 222 is apparent.

$$\begin{array}{r} 851 \\ 666 \\ \hline 185 \end{array} \qquad 222$$

It will be easy to follow this out through the successive steps:

$$\begin{array}{cc} 185 & 222 \\ & \underline{185} \\ & 37 \\ \underline{} & \\ 185 & 37 \end{array}$$

37 divides 185 exactly, and is thus the greatest common measure of 185 and 37; so that 37 is the greatest common measure of

$$851 \text{ and } 1073.$$

The class will now see that

$$851 = 23 \times 37$$
$$1073 = 29 \times 37$$

and a conviction will be added to the proof. Then the identity of the work with the following may be shown:

$$\begin{array}{r} 851)\overline{1073}(1 \\ \underline{851} \\ 222)\overline{851}(3 \\ \underline{666} \\ 185)\overline{222}(1 \\ \underline{185} \\ 37)\overline{185}(5 \\ 185 \end{array}$$

We see now that a definite method has been evolved, and when the class has been exercised in applying it, it may be well to explain certain artifices by means of which the work may be shortened, or exhibited in a neater form. For example, the work of finding the greatest common measure of 851 and 1073, as given above, may be presented as follows:

	1	3	1	5	
1073	851	222	185	37	= G. C. M.
851	666	185	185		
222	185	37			

Or the work might be conveniently arranged as in the following example: Find the greatest common measure of 158938 and 531206.

		531206
158938	3	476814
108784	2	54392
50154	1	50154
46618	11	4238
3536	1	3536
3510	5	702
26	27	702

The quotients appear in the middle column, and the work explains itself.

It is to be observed that if any common factor is easily discoverable in the two given quantities, it is better first to divide both quantities by the common factor. If, also, a prime factor is found in only one of the quantities which are in operation for the greatest common measure, it may be struck out. In the last example, for instance, the first remainder is divisible by 8, while the corresponding number on the other side (the first divisor) is divisible by 2. We may therefore divide this number by 2 and the other by 8, reserving **2** as part of the required common measure. These factors being removed, we operate with the quotients, 79469 and 6799. The latter divides the former with remainder 4680; this, it is obvious, has the factors 40, 13, 9. Hence, if the two original quantities have a common factor, it is 13 × 2—a result obtained by the actual work.

This study of the measures of numbers suggests classifications of numbers. Numbers may be (1) even or odd, according as they do or do not contain 2 as a factor; (2) composite or prime, according as they are or are not resolvable into simpler factors.

Two numbers may have no factors in common, though each of them may be composite; they are then said to be *prime to each other*. It will be supposed that the class is familiar with these classifications and definitions before proceeding to a study of least common multiple.

LEAST COMMON MULTIPLE.

In the presentation of least common multiple, it is necessary—as indeed it always is in the introduction of a new process—first to bring out clearly the essential facts and ideas upon which the process rests. Here the pupil must first get a clear idea of the terms multiple, common multiple, least common multiple. A factor (or measure) of 15 is 3; 15 is called a *multiple* of 3; it represents the quantity that 3 exactly measures. The pupil will now be asked to name different multiples of 3, say, and will see that he may name or write down as many as he chooses. Then, if a series of multiples of 2 be written down, so that we have the two series:

3, 6, 9, 12, 15, 18, 21, 24, . . .
2, 4, 6, 8, 10, 12, 14, 16, 18, 20, 22, 24, . . .

he will see that there are numbers which are at the same time multiples of 2 and 3, and are therefore *common* multiples of 2 and 3. These common multiples are, here, 6, 12, 18, . . . Then, because we have started with the smallest multiples of the numbers, 6 is the smallest or *least* common multiple of 2 and 3. At this point the pupil can hardly fail to see that the second common multiple is $6+6$, the third is $6+6+6$, etc.; in other words, that all the common multiples of two numbers are formed by repeating as an addend the *least* common multiple. He can then be led to see the reason for this, viz., that (referring to the foregoing example), in order to get a common multiple of 2 and 3 larger than 6, it will be necessary to add to 6 a common multiple of 2 and 3, so that 6 is the least number that can be used.

Numbers Prime to Each Other.—A necessary step preliminary to teaching the formal process is the bringing out of the fact that the least common multiple of two numbers prime to each other is their product. For example, take the numbers 5 and 7: a common multiple must have 5 as a factor and 7 as a factor; it is, therefore, 5 multiplied by another factor. But since the multiple contains 7, and since 7 is prime to 5, the other factor of the multiple must contain 7. Hence, since the smallest multiple of 7 is 7, the least common multiple of 5 and 7 is 5×7. Similarly, a common multiple of 4 and 9 is a multiple of 4, and is therefore 4 multiplied by another factor; but the common multiple is a multiple of 9, and therefore, since 4 contains no factor of 9, the other factor must contain 9; thus, since 9 is the least number that will contain 9, the least common multiple of 4 and 9 is 4×9. This fact has generally been taken as self-evident. It seems best, however, to help the pupils to get full possession of it, and the somewhat abstract discussion should be illustrated by a series of multiples of the numbers

$$5, 10, 15, 20, 25, 30, 35,$$
$$7, 14, 21, 28, 35, 42, 49.$$

Here 30, for example, can not be a multiple of 7, being made up of the factors 5, 2, and 3, all of which are prime to 7.

The next step will be to find the least common multiple of two numbers not prime to each other, say 36 and 48. Resolving each of these into simple factors, we see that

$$36 = 2 \times 2 \times 3 \times 3$$
$$48 = 2 \times 2 \times 2 \times 2 \times 3$$

Thus a common multiple must have 2 as a factor four times, and 3 as a factor two times, so that the least common multiple is

$$2 \times 2 \times 2 \times 2 \times 3 \times 3, \text{ or } \mathbf{144}$$

It will now be easy to bring out the relation of this to the following process:

$$
\begin{array}{r}
2)\overline{36, \ 48} \\
2)\overline{18, \ 24} \\
3)\overline{9, \ 12} \\
3, \ 4
\end{array}
$$

The first division shows that $36 = 2 \times 18$, and $48 = 2 \times 24$. Thus a common multiple must have 2 as a factor, and its remaining factors must make up a number which will contain exactly the numbers 18 and 24. Similarly, a common multiple of 18 and 24 must have the factor 2, and its remaining factors must make up a number which will contain exactly the numbers 9 and 12, etc. Clearly, then, a common multiple must have the factors $2 \times 2 \times 3$, and its remaining factors must constitute a number that will contain exactly 3 and 4, two numbers prime to each other, and having therefore 4×3 for *least* common multiple. Hence the least common multiple of the given numbers is

$$2 \times 2 \times 3 \times 3 \times 4, \text{ or } \mathbf{144}$$

It will now be easily seen that it would have been allowable to divide at once by **12**, the greatest number that will divide each of 36 and 48, and the pupil can deduce a definite method of finding the least common multiple of any two given numbers, whether he can discover factors at once, or is compelled to resort to the process of finding the greatest common measure of the two numbers.

The process may now be extended to the case of three or more numbers:

$$\begin{array}{r|rrr} 2 & 12, & 15, & 18 \\ \hline 3 & 6, & 15, & 9 \\ \hline & 2, & 5, & 3 \end{array}$$

The least common multiple must have the factor 2, and must further provide for the numbers 6, 15, 9; to do this, an additional factor, 3, must be introduced, and it remains yet to provide for the numbers 2, 5, 3. These latter are prime to one another, so that the least number that will contain them is $2 \times 5 \times 3$; and hence the least common multiple of 12, 15, 18 is

$$2 \times 3 \times 2 \times 5 \times 3, \text{ or } 180$$

The preceding example might have been treated as follows:

$$\begin{array}{r|lll} 6 & 12, & 15, & 18 \\ \hline 3 & 2, & 15, & 3 \\ \hline & 2, & 5, & 1 \end{array}$$

and least common multiple $= 6 \times 3 \times 2 \times 5 = 180$. Here the first division was by 6, which is seen to divide 12 and 18. But the division by a composite number *may* lead to a multiple which is not the least.

Example:

$$\begin{array}{r|lll} 3 & 6, & 15, & 42 \\ \hline 2 & 2, & 5, & 14 \\ \hline & 1, & 5, & 7 \end{array}$$

$\therefore$ L. C. M. $= 3 \times 2 \times 5 \times 7 = \mathbf{210}$

If 6 had been taken as the first divisor, the work would have stood thus:

$$\begin{array}{r|lll} 6 & 6, & 15, & 42 \\ \hline & 1, & 15, & 7 \end{array}$$

and the result would have been

$$6 \times 15 \times 7 = 3 \times 2 \times 3 \times 5 \times 7 = 630$$

The reason for the difference in results is apparent. In the latter process, after the factor 6 of the least common multiple was found, it was supposed that there was still need to provide for all the factors of 15—i. e., for 3 and 5; whereas the factor 6, already taken, contains the factor 3. It will also be seen why, in the case of finding the least common multiple of 12, 15, and 18, a similar way of proceeding gave the correct result. The pupil is now led to see that in finding the least common multiple of several numbers it is frequently necessary, and therefore (practically) always advisable, to divide out by factors that are prime numbers.

Larger Numbers.—The thought which from the first has directed this presentation, and which is now clearly understood by the pupils is, the least common multiple of several numbers must contain (1) all the *different* factors in the numbers, (2) each in the highest power it has in any of the numbers. The pupils will now easily deduce the rule for applying the greatest common measure to find the least common multiple of numbers that can not be resolved by inspection. If the least common multiple of 851 and 1073 is required, they will find, as before, $851 = 23 \times 37$, $1073 = 29 \times 37$, and will see at once that therefore the least common multiple is $23 \times 29 \times 37$. On comparing this with the two numbers and their greatest common measure they will readily see that $23 = \frac{23 \times 37}{37} = \frac{851}{37}$; and that $29 = \frac{29 \times 37}{37} = \frac{1071}{37}$; and that therefore the least common multiple of two numbers may be found by dividing *either* of them by their greatest common measure and multiplying the quotient by the other. This may, of course, be readily applied to all cases; for example, find the least common multiple of 12, 15, 18. The greatest common measure of 12×15 is 4; ∴ their least common multiple is $4 \times 15 = 60$; the greatest common measure of 60 and 18 is 6; ∴ the least com-

17

mon multiple is $\frac{60}{6} \times 18 = 180$. Similarly, find the least common multiple of 12, 20, 36, 60, 54.

The least common multiple of 12 and 20 is   60.
" " " 60 and 36 is 180.
" " " 180 and 54 is 540.

Speaking from actual experience, we have not a doubt that pupils can be led to the statement of the principles and rules in general forms: If $a = mc$, and $b = nc$, when c is the greatest common measure of two quantities a and b, then the least common multiple of a and b is mnc, which is $\frac{mc}{c} \times nc$, or $\frac{nc}{c} \times mc$, etc.

In closing this chapter, it is scarcely necessary to emphasize the importance of adding an interest to this part of the work in arithmetic by drawing upon the great variety of practical problems as illustrating measures and multiples; or to speak of the need of having the pupils give the reasons which led them to conclude that the problem was concerned with the finding of a measure or a multiple, as the case may be.

CHAPTER XIII.

FRACTIONS.

AFTER the previous discussion on the nature of fractions (see especially Chapter VII) and their psychological relation with the fundamental operations, a brief reference to some of the points brought out is all that is needed as an introduction to the formal teaching of fractions.

Number depends upon measurement of quantity. This measurement begins with the use of inexact units—the counting of like things—and gives rise to addition and subtraction. From this first crude measurement is evolved the higher stage in which exactly defined units of measure are used, and in which multiplication, division, and fractions arise. Multiplication and division bring out more clearly the idea of number as measurement of quantity—as denoting, that is, (*i*) a unit of measure and (*ii*) times of its repetition. The fraction carries the development of the measuring idea a step further. As a mental process it constitutes a more definite measurement by consciously using a defined unit of measure; and as a *notation*, it gives complete expression to this more definite mental process. Fractions therefore employ more explicitly both the conceptions involved in multiplication and division—namely, analysis of a whole into exact units, and synthesis of these into a defined whole. The idea of fractions is present from the first, because division and multiplication are implied from the first. There is no number without measurement, nor measurement without fractions. Even in whole numbers, as has been pointed out, both "terms" of a fraction are implied in the accurate interpretation of the measured quantities.

Since there is nothing new in the process of fractions, so in the teaching of fractions there is nothing essentially different from the familiar operations with whole numbers. If the idea of number as measurement has been made the basis of method in primary work and in the fundamental operations, the fraction idea must have been constantly used, and there is absolutely no break when the pupil comes to the formal study of fractions. There is only before him the easy task of examining somewhat more attentively the nature of the processes he has long been using. The suggestions made in reference to primary teaching and formal instruction in the fundamental operations apply with equal force to the teaching of fractions. The measuring idea is to be kept prominent: avoidance of the fixed unit fallacy and its logical outgrowth, the use of the undefined qualitative unit—the pie and apple method—as the basis for developing "fractional" units of measure, and the "properties of fractions"; the essential property of the *unit* in measurement—the *measured* part of a *measured* whole; the logical and psychological relation between the *number* that defines the measuring *unit* and the *number* that defines the measured *quantity;* or as it is sometimes expressed, the "relation between the *size* of the parts" (the measuring units) and the *number* of the parts composing or equalling the measured quantity; these and all kindred points that have been brought out in discussing number as measurement, and numerical operations as simply phases in the development of the measuring idea, can not be ignored in the teaching of fractions, because they can not be ignored in the teaching of whole numbers. Exact number demands definition of the unit of measure; the fraction completely satisfies this demand by stating or defining expressly the unit of measure. In all number as representing measured quantity the questions are: What is the unit of measure, and how is it defined or measured? How many units equal or constitute the quantity? These questions only number in its fractional form completely answers. It is the completion of the psychical process of number as measurement of quantity; the idea of the quantity is made definite, and it is definitely expressed.

While the following treatment of fractions is in strict line with the principles of number set forth in these pages, and has stood the test of actual experience, it is given only by way of suggestion. The principles are universal and necessary; devices for their effective application are within certain limits individual and contingent. Principles are determined by philosophy, devices by rational experience. The teacher must be loyal to principles, but the slave of no man's devices.

I. THE FUNCTION OF THE FRACTION.

1. In its primary conception a fraction may be considered as a number in which the unit of measure is

expressly defined. In the quantities 4 dimes, 5 inches, 9 ounces, the units of measure are not explicitly defined; their value is, however, implied, or else there is not a definite conception of the quantity. In $\frac{4}{10}$, $\frac{5}{12}$ foot, $\frac{9}{16}$ pound, the units of measure are explicitly defined; and each of these expressions denotes four things: 1. The unity (or standard) of reference from which the actual unit of measure is derived. 2. How this unit is derived from the unity of reference. 3. The absolute number of these derived units in the quantity. 4. This number is the ratio of the given quantity to the unity of reference.

For example, in $\frac{9}{16}$ pound, the unity of reference is one pound; it is divided into sixteen equal parts, to give the direct measuring unit; the number of these units in the given quantity is nine; the ratio of the given quantity to the unity of reference is nine.

2. If properly taught, the pupil knows—if not, he must be made to know—that any quantity can be divided into 2, 3, 4, 5 . . . n equal parts, and can be expressed in the forms $\frac{2}{2}$, $\frac{3}{3}$, $\frac{4}{4}$, $\frac{5}{5}$ . . . $\frac{n}{n}$. Familiar with the ideas of division and multiplication which become explicit in fractions, he learns in a few minutes (has already learned, if he has been rationally instructed) that any quantity may be measured by 2 halves, or 3 thirds, or 4 fourths . . . or n $n$ths; that to take a half, a third, a fourth . . . an nth of any quantity, it is only necessary to divide by 2, or 3, or 4 . . . or n; that if, for example, 16 cents, or 16 feet, or 16 pounds has been divided into four parts, the counts of the units in each case are one, two, three, four, or one fourth, two fourths, three fourths, four fourths; that each of these units—

fourths—is measured by other units, and can be expressed as integers, namely, 4 cents, 4 feet, 4 pounds, and so on, with kindred ideas and operations.

3. *The Primary Practical Principle in Fractions.*—It is clear that this complete expression for the number process is the fundamental principle employed in the treatment of fractions: if both terms of a fraction be multiplied or divided by the same number, the numerical value of the fraction will not be changed. This principle is usually "demonstrated"; it is, however, involved in the very conception of number, and seems as difficult to demonstrate as the definition of a triangle; but intuitions and illustrations to any extent may be given. Any 12-unit quantity, for example, is measured by $\frac{2}{2}$, $\frac{4}{4}$, $\frac{12}{12}$, or by $\frac{3}{3}$, $\frac{6}{6}$, $\frac{12}{12}$; the identity of the quantity remains unchanged in the changing measurements. Moreover, if half the quantity be measured, the identity of $\frac{1}{2}$, $\frac{2}{4}$, $\frac{3}{6}$, $\frac{6}{12}$ is seen at once. The principle is, of course, that in a given measured quantity the "size" of the units varies inversely with their number. This principle is said to be beyond the comprehension of the pupil. On the contrary, if constructive exercises, such as have been described, have been practised, there comes in good time a complete recognition of the principle. When, for instance, the child measures off any 24-unit quantity by twos, threes, fours, sixes, eights, he can not help feeling the relation between the magnitude and the number of the measuring parts. This is, in fact, the process of number.

Proof of the Principle.—If the first vague awareness of the relation does not grow into a clear comprehension of it, clearly the method is at fault. In any case, if the pupil does not understand the principle after rationally using it, any formal "demonstration" is a mere delusion; for any so-called demonstration is grounded on the principle—in general *is* the principle—merely illustrated or used in a disguised form. For example: Prove $\$\frac{3}{4} = \$\frac{15}{20}$. Since $\$\frac{3}{4} = \$\frac{1}{4} \times 3$: multiply by 4, and we have $\$\frac{3}{4} \times 4 = \$\frac{1}{4} \times 4 \times 3 = \$3$; multiply these equals by 5:

$\therefore \$\frac{3}{4} \times 4 \times 5 = \$3 \times 5 = \$15$; but also $\$\frac{15}{20} \times 20 = \$15$;

$\therefore \$\frac{3}{4} \times 20 = \$\frac{15}{20} \times 20$; dividing equals by equals;

$\therefore \$\frac{3}{4} = \$\frac{15}{20}$. Or, generally: $\frac{a}{b} \times b = a$; multiply both sides by n.

$\therefore \frac{a}{b} \times nb = na$; but $\frac{na}{nb} \times nb = na$;

$\therefore \frac{a}{b} \times nb = \frac{na}{nb} \times nb$; dividing both sides by nb;

$\therefore \frac{a}{b} = \frac{na}{nb}$.

These and similar proofs are in essence the idea already considered: that if a quantity is divided into a certain number of equal parts, each part has a certain value; if into twice the number of parts, each part has *half* the value of the former part; if into three times as many parts, each has a third of the value; if into n times as many parts, each has 1 nth of the value.

If formal proof is wanted of this important principle (which is, once more, the principle of number), the following is perhaps as intelligible as any other. To prove, for example, that $\$\frac{3}{4} = \$\frac{27}{36} = \$\frac{75}{100} =$, etc., we have $\$4 = \$\frac{36}{36} = \$\frac{100}{100}, =$, etc.;

$\therefore \$\frac{1}{4} = \$\frac{9}{36} = \$\frac{25}{100} =$, etc.

$\therefore \$\frac{3}{4} = \$\frac{27}{36} = \$\frac{75}{100} =$, etc.

FRACTIONS.

4. *The Fraction as Division.*—While in its primary conception the fraction is not simply a formal division, it nevertheless involves the idea of division, and can not be fully treated without identifying it with the formal process. The quantity $\frac{7}{12}$ foot, first regarded as $\frac{1}{12}$ foot × 7, must be recognized in its psychological correlate, 7 feet × $\frac{1}{12}$—i. e., 7 feet ÷ 12. As has been shown more than once, these measurements can not but be recognised as two phases of the same measurement, whenever the process becomes the object of conscious attention. It is the law of commutation, the connection between the number and the magnitude of the units in a measured quantity. If we do not know that $\frac{1}{3}$ of 3 times a quantity is 3 times $\frac{1}{3}$ of the quantity—or, generally, that $\frac{1}{n}$ of n times $q = n$ times $\frac{1}{n}$ of q—we have no clear conception of number. If a quantity is measured by $\frac{1}{12}$ of a certain unity of reference taken 7 times, this is seen to be identical with $\frac{1}{12}$ of one of these unities $+ \frac{1}{12}$ of a second $+ \frac{1}{12}$ of a third . . .; that is, in all, $\frac{1}{12}$ of seven of them.

Numerous illustrations and so-called proofs may be given. Examples:

(1.) Show that $\frac{3}{4}$ of any quantity is equal to $\frac{1}{4}$ of 3 times the quantity. Let A B be any quantity whatever, measured in fourths and expressed as $\frac{4}{4}$; and C D, E F, each measured quantities equal to A B. It is obvious that A K, which is $\frac{3}{4}$ of A B, is equal to A G + C H + E L; that is, equal to $\frac{1}{4}$ of 3 times A B.

(2) To show that $\frac{1}{12}$ foot × 7 = 7 feet × $\frac{1}{12}$—i. e.,

= 7 feet ÷ 12. The unit of reference, 1 foot, may be thought of and expressed as 12 twelfths foot:

∴ 7 feet = 84 twelfths foot;
∴ 7 feet × $\frac{1}{12}$ = 7 twelfths foot = $\frac{1}{12}$ foot × 7.

(3) To prove that $\$\frac{1}{4} \times 3$ is equal to $\$3 \times \frac{1}{4}$:

4 times $\$\frac{1}{4} = \$1$; multiply these equals by 3;
∴ 4 times $\$\frac{1}{4} \times 3 = \$3$;
∴ $\$\frac{1}{4} \times 3 = \$3 \div 4$. Or, using q for any quantity, 4 times $\frac{1}{4}q = q$;
∴ 4 times $\frac{1}{4}q \times 3 = 3q$; hence, $\frac{1}{4}q \times 3 = 3q \div 4$.

(4) Or, generally, $\frac{1}{n}q \times m = mq \div n$. For

n times $\frac{1}{n}q = q$;
∴ n times $\frac{1}{n}q \times m = mq$;
∴ $\frac{1}{n}q \times m = mq \div n$.

Such formal proofs are useful and even necessary, but are likely to be misleading unless the pupil has evolved, from rational *use* of the principle, a clear idea of the relation between times and parts, the importance of which has been emphasized in this book; he is apt to become a mere spectator in the manipulation of symbols, rather than a conscious actor in the mental movement which leads to complete possession of the thought.

II. Change of Form in Fractions.

1. From what has been already said, it appears that any quantity may be expressed in the form of a fraction having any required denominator. Express 9 yards in eighths of a yard. Since the unit of measure is $\frac{8}{8}$, 9

such units is $\frac{72}{8}$. Similarly, $7 expressed as hundredths is $\frac{700}{100}$, etc. In general, any quantity of q units of measure expressed as $n$ths is $\frac{nq}{n}$.

2. In the same way, any quantity expressed in fractional form may be changed to an equivalent fraction having any denominator. Transform $\$\frac{4}{5}$ into an equivalent fraction having denominator 20. We can follow either of two plans:

(1) 20 is a multiple of 5 by 4; we therefore multiply both terms of the given fraction by 4, getting $\$\frac{16}{20}$. This is best in practice.

(2) Since the new denominator is to be 20, we regard the unit of measure as $\$\frac{20}{20}$; $\$\frac{4}{5}$ is $\$\frac{1}{5} \times 4$; but $\$\frac{1}{5} =$ one fifth of $\$\frac{20}{20} = \$\frac{4}{20}$;
$\therefore \$\frac{4}{5}$ $= \$\frac{16}{20}$.

It may be remarked that in such transformations the new denominator is generally a multiple of the original denominator. If it is not, the new equivalent fraction will be complex, it will have a fractional numerator. Thus, if it is required to transform $\frac{4}{8}$ yard to an equivalent fraction with denominator 12, we multiply both terms by $1\frac{1}{3}$, with the result $\frac{5\frac{1}{3}}{12}$.

3. It is often necessary or convenient to reduce a fraction to its lowest terms—that is, to express it in terms of the largest unit of measure as defined by the unity of reference. This is done by dividing both terms of the fraction by their greatest common measure; thus, $\$\frac{6}{12}$ is equivalent to $\$\frac{1}{2}$, in which the quantity is expressed in the largest unit of measure, as defined by the unity of reference, the dollar. The principle in-

volved is that stated in I, 3—viz., the numerical *value* of the units is increased a number of times, the *number* of them is diminished the same number of times.

In practice, the greatest common measure can generally be found by inspection, as described in Chapter XII. Thus, $\frac{612}{684} = \frac{2 \times 2 \times 3 \times 3 \times 17}{2 \times 2 \times 3 \times 3 \times 19} = \frac{17}{19}$. In some cases the greatest common measure must be found by the general method described in the same chapter. Thus, if the proposed fraction is $\frac{79469}{265603}$, we should discover the greatest common divisor to be 13; on dividing both terms of the given fraction there results $\frac{6113}{20431}$, which is the simplest of all equivalent fractions.

4. In changing a mixed number to an improper fraction, and *vice versa*, the primary principle of fractions applies at once:

(1) Reduce $75\frac{2}{3}$ yards to an improper fraction. The expression $= 75$ yards $+ \frac{2}{3}$ yard; express 75 yards in form of a fraction with denominator 3:

1 yard is $\frac{3}{3}$ yard;
75 yards is $\frac{3}{3}$ yard $\times 75 = \frac{225}{3}$ yards;
$\therefore$ 75 yards $+ \frac{2}{3}$ yard $= \frac{225+2}{3}$ yards $= \frac{227}{3}$ yards.

(2) In the converse operation either consider the problem as a case of formal division giving $75\frac{2}{3}$, or consider the expression as denoting so many *thirds* of a yard; then 3-thirds $=$ *one* yard; how many 3-thirds in 227 thirds? Evidently, as before, a case of division, giving 75 ones and two-thirds remainder—that is, $75\frac{2}{3}$ yards. It may be observed that in (1), while the primary measurement of the quantity is 3 units $\times 75 + 2$

FRACTIONS.   251

units, and we multiply 3 by 75, it is equally logical to use the correlate 75 units $\times$ 3. (See page 77.)

5. *Comparison of Fractions. Common Denominator.*—It is often necessary to transform fractions having different denominators to equivalent fractions with the same denominator. Quantities can not be definitely compared, we have seen, unless they have the same unit of measure. We can not compare directly 5 feet and 5 rods; they must both be expressed in terms of the same measuring unit. So with 4 dollars and 7 dimes. In like manner, $\$\frac{3}{4}$ and $\$\frac{4}{5}$ can not be definitely compared; they have a common (primary) unit of reference, but not a common (actual) unit of measure; but they can both be expressed in terms of a common unit of measure. They can be expressed as $\$\frac{15}{20}$, $\$\frac{16}{20}$, or $\$\frac{30}{40}$, $\$\frac{32}{40}$, etc. For comparison, it is generally most convenient to express such quantities in terms of the *greatest* common unit of measure, and this is determined by the *least* common divisor or denominator.

(1) Compare the quantities $\frac{5}{6}$ foot and $\frac{6}{7}$ foot. Here the least common multiple of the denominators is 42, and the fractions become $\frac{35}{42}$ foot and $\frac{36}{42}$ foot; the latter fraction is therefore the greater.

(2) Which is the greater, $\$\frac{15}{32}$ or $\$\frac{59}{128}$? The least common denominator is 128, and the 128th of a dollar is then the common unit of measure; the fractions are therefore $\$\frac{15 \times 4}{128}$, $\$\frac{59}{128}$, and hence the former is the greater.

It is to be observed that if the question is simply, Which is the greatest of a number of fractions? it can be answered by reducing them, by a similar process, to a common numerator.

(1) Which is the greater, $\tfrac{5}{6}$ or $\tfrac{4}{7}$? The least common multiple of the numerators is 30; multiplying both terms of the first fraction by 6, and of the second by 5, we have $\tfrac{30}{36}$, $\tfrac{30}{35}$; and since the latter has the smaller denominator (therefore greater unit of measure), it is the larger fraction.

(2) Compare the quantities $\tfrac{3}{4}$, $\tfrac{7}{12}$, $\tfrac{5}{8}$, by reducing them to a common numerator. The least common multiple of 3, 7, 5 is 105; the multipliers for the terms of the respective fractions are, therefore, 35, 15, 21, giving $\tfrac{105}{140}$, $\tfrac{105}{180}$, $\tfrac{105}{168}$; hence, the first fraction is the greatest, and the second one the least. In this case the comparison would be easier by reducing to a common denominator.

The Greatest Common Measure and Least Common Multiple of Fractions.—Here, also, there is nothing essentially different from the corresponding operations in whole numbers. As has been said, quantities, in order to be compared, must be expressed in terms of the same unit of measure. If fractions have a common denominator—represent, that is, quantities defined by the same unit of measure—the ordinary rules for measures and multiples at once apply. Examples:

(1) Find the greatest common measure of $\tfrac{5}{6}$ yard and $\tfrac{4}{9}$ yard. Expressed with the same denominator these become $\tfrac{30}{36}$ yard, $\tfrac{16}{36}$ yard. The greatest common measure of 30 and 16 is 2; therefore $\tfrac{2}{36}$ is the greatest common measure required.

(2) What is the greatest length that is contained an integral number of times in $18\tfrac{2}{5}$ feet and $57\tfrac{1}{2}$ feet? Change to improper fractions with least common denominator: $\tfrac{184}{10}$, $\tfrac{575}{10}$. The greatest common measure

of these numerators is 23; therefore $\frac{23}{10}$, or $2\frac{3}{10}$, is the required number.

(3) Four bells begin tolling together; they toll at intervals of 1, $1\frac{1}{8}$, $1\frac{1}{12}$, $1\frac{3}{10}$ seconds respectively; after what interval will they toll together again?

Here the least common multiple of the numbers is required. Change to improper fractions with least common denominator: $\frac{120}{120}$, $\frac{135}{120}$, $\frac{130}{120}$, $\frac{156}{120}$. The least common multiple of the numbers is 14,040, and the least common multiple required is therefore $\frac{14040}{120} = 117$; therefore the required interval is 117 seconds.

III. THE FUNDAMENTAL OPERATIONS.

1. *Addition of Fractions.*—(1) When the fractions have a common denominator—that is, when they denote the same unit of measure—the process is the same as in whole numbers. There is no essential difference between the operations 3 dimes $+$ 4 dimes and $\$\frac{3}{10} + \$\frac{4}{10}$. The only difference is in the mode of *expressing* the unit of measure—seven dimes in the one case, seven tenths of a dollar in the other.

(2) When the denominators are different the fractions must be reduced to equivalent fractions having a common denominator—that is, they must be expressed in terms of a common unit of measure (see II, 6). We can not add $4 and 4 half-eagles, nor 4 feet and 4 yards, till we express the quantities in a common unit of measure, the first two in the form $4 + $20, and the second in the form 4 feet + 12 feet. So we can not add $\frac{3}{4}$ yard and $\frac{4}{5}$ yard without first expressing them in terms of a common measuring unit. For convenience we select, as before said, the *greatest* common unit of meas-

ure as defined in relation to the primary unit of reference, the yard. This greatest unit is given by the least common multiple of the denominators 4 and 9, which is 36. The quantities expressed in the common *unit* of measure are therefore $\frac{27}{36}$ yard and $\frac{16}{36}$ yard, and their sum is $(27+16=43)$ 36ths of a yard, or $\frac{43}{36}$ yard. The operation is essentially the same as that of finding the sum $(27+16)$ inches.

(3) In addition of mixed numbers it is best to find the sum of the whole numbers, the sum of the fractions, and then the sum of these two results. Improper fractions should, in general, be expressed as mixed numbers. Find the sum of $13\frac{4}{9}$ yards, $17\frac{5}{6}$ yards, $3\frac{3}{4}$ yards. The sum is

$$13\tfrac{4}{9} + 17\tfrac{5}{6} + 3\tfrac{3}{4} = 13 + 17 + 3 + \tfrac{4}{9} + \tfrac{5}{6} + \tfrac{3}{4}$$
$$= 33 + \tfrac{16+30+27}{36}$$
$$= 33 + 2\tfrac{1}{36} = 35\tfrac{1}{36} \text{ yards.}$$

2. *Subtraction of Fractions.*—Since subtraction is the inverse of addition, the same principles and methods apply in both. In subtraction of mixed numbers it is generally best first to change the fractional parts to equivalent fractions with the same unit of measure, and then perform the subtraction. Example:

(1) How much was left of $\$10\tfrac{7}{10}$ after a payment of $\$5\tfrac{2}{5}$?

Expressing the fractional parts with common denominator:

$$\$10\tfrac{7}{10} - \$5\tfrac{4}{10} = \$5\tfrac{3}{10}.$$

(2) How much was left of $16\tfrac{2}{3}$ yards of cloth after $6\tfrac{3}{5}$ yards were cut from it?

Reducing the fractional parts to common denominator:

FRACTIONS. 255

$$16\tfrac{3}{8} = 16\tfrac{27}{72} = 15\tfrac{72+27}{72}$$
$$6\tfrac{8}{9} = 6\tfrac{64}{72}$$

$$15\tfrac{72+27}{72}$$
$$-6\tfrac{64}{72}$$
$$= 9\tfrac{35}{72} \text{ yards remainder.}$$

Here the fractional part of the subtrahend is greater than the fractional part of the minuend; the minuend is therefore changed to the form $15 + 1 + \tfrac{27}{72} = 15\tfrac{72+27}{72} = 15\tfrac{99}{72}$. In actual work we may take 64 from 72, the denominator of the minuend fraction, and add to the remainder the numerator of the minuend fraction. Thus, we can not subtract 64 from 27; we subtract it from 72, and add the remainder, 8, to 27, getting $\tfrac{35}{72}$. This is equivalent to taking $1 (= \tfrac{72}{72})$ from the minuend, and uniting it with the minuend fraction, as has been done in the example.

3. *Multiplication of Fractions.* — (1) When the multiplicand is a fraction and the multiplier a whole number, the operation is exactly like multiplication of integers. To find the cost of 12 yards of cloth at $\$\tfrac{3}{4}$ a yard, we multiply 3 by 12 and define the product by the proper unit of measure. In finding the cost of 12 yards at $3 a yard, the complete process is $\$1 \times 3 \times 12$; we operate with the pure numbers 3 and 12, getting 36, and define the product by naming the proper unit of measure *one dollar;* the cost is then 36 dollars. In the proposed case we do exactly the same thing: $\$\tfrac{1}{4} \times 3 \times 12$—that is, 36 times the proper unit of measure ($\$\tfrac{1}{4}$)—and the product is 36 *quarter dollars.* Neither the process nor the product changes, because the *unit*, or the manner of writing it, happens to change.

(2) When the multiplier is a fraction, exactly the

same principles hold; in fact, the measured quantity, $\$\tfrac{3}{4} \times 12$, is identical with $\$12 \times \tfrac{3}{4}$. In this conception of quantity (money value) we have nothing to do with yards, and either form of the measurement may be taken. In fact, $\$\tfrac{3}{4} \times 12$ is $\tfrac{3}{4}$ of $\$1 + \tfrac{3}{4}$ of $\$1 + \tfrac{3}{4}$ of $\$1$, and so on 12 times—that is, $\tfrac{3}{4}$ of $\$12$. The multiplicand is *always* a unit of measure; the multiplier always shows how this unit is treated to make up the measured whole. It is purely an operation. In this example the denominator shows how the unit $\$12$ is to be dealt with in order to yield the derived unit of measure: it is to be divided into *four* parts, and the derived unit thus found is to be taken three times. As already shown, from the nature of the fraction it denotes three times one fourth of the multiplicand, or one fourth of three times the multiplicand—that is, $\$\tfrac{12}{4} \times 3$, or $\$\tfrac{12 \times 3}{4}$.

(3) The explanation usually given of the process is in harmony with this. This explanation considers the multiplicand as a case of pure division; that is, $\tfrac{3}{4}$ is one fourth of 3, and to multiply a quantity by $\tfrac{3}{4}$ is to take one fourth of 3 times the quantity. In fact, in all operations with fractions the idea of division, as well as of multiplication, is present; a factor and a divisor are always elements in the problem.

(4) The method to be followed when both factors are of fractional form involves nothing different from the other two cases.

The price of $\tfrac{4}{9}$ yard of cloth at $\$\tfrac{3}{4}$ a yard is to be found. The result is indicated by $\$\tfrac{3}{4} \times \tfrac{4}{9}$; that is, as before, 4 times a certain quantity is to be divided by 9, or $\tfrac{1}{9}$ of the quantity is to be multiplied by 4. In the first case, $\$\tfrac{3}{4} \times 4$ is *either* $\$\tfrac{1}{4} \times 3 \times 4 = \$\tfrac{1}{4} \times 12 =$

$3; or $\frac{3}{4} \times \frac{4}{9} \times 3 = $1 \times 3 = 3; and $\frac{1}{9}$ of this is $\frac{3}{9}$, or $\frac{1}{3}$.

It may be observed that we may change the multiplicand into an equivalent fraction with a unit of measure determined by a multiple of the denominators. In $\frac{3}{4} \times \frac{4}{9}$, for example, we have $\frac{27}{36} \times \frac{4}{9} = \frac{3}{36} \times 4 = \frac{3}{9} = \frac{1}{3}$. The complete process is seen to be $\frac{3}{4} \times \frac{4}{9} = \frac{12}{36}$. But since numerators are always factors of a dividend, and denominators factors of a divisor, common factors may be divided out. In $\frac{3}{36} \times 4$, for instance, the value of the quantity is the same whether we take 4 times the number of units ($= \frac{12}{36}$), or make the units 4 times as large ($\frac{3}{9}$). This is nothing but the application of fundamental principles (see page 226) of multiplication and division. If we have to divide 210 by 21, we may proceed thus: $\frac{210}{21} = \frac{3 \times 7 \times 5}{3 \times 7} = \frac{3}{3} \times \frac{7}{7} \times 5 = 5$. Again, $32340 \div 385 = \frac{7 \times 7 \times 11 \times 2 \times 6 \times 5}{11 \times 5 \times 7} = \frac{7}{7} \times \frac{11}{11} \times \frac{5}{5} \times 7 \times 2 \times 6 = 1 \times 84 = 84$.

4. *Division of Fractions.*—(1) When the divisor is an integer and the dividend a fraction.

Paid $\frac{7}{10}$ for 5 yards of calico, what was the price per yard? One yard will cost one fifth of $\frac{7}{10}$, or $\frac{7}{50}$. At $5 a yard, how much lace can be bought for $\frac{7}{10}$? The answer is indicated in $\frac{7}{10} \div 5; the quantities must have the same unit of measure, and the expression is equivalent to $\frac{7}{10} \div \frac{50}{10} = 7 \div 50 = \frac{7}{50}$; hence, $\frac{7}{50}$ yard of lace can be bought.

(2) When the divisor is a fraction, and the dividend an integer.

At $\frac{3}{4}$ a yard, how many yards of dress goods can be bought for $6?

The number of yards is given in $6 \div $\frac{3}{4}$, where, again, the quantities must be reduced to the same unit of measure: $\$6 \div \$\tfrac{3}{4} = \$\tfrac{24}{4} \div \tfrac{3}{4} = 24 \div 3 = 8$; hence, 8 yards can be bought.

Paid $6 for $\tfrac{3}{4}$ yard of velvet, what was the price per yard? The cost is given in $\$6 \div \tfrac{3}{4}$, which means that $\tfrac{1}{4}$ of 3 times the quantity sought is $6, and therefore it is $\$6 \times 4 \div 3 = \$8$. Or, by the law of commutation, $\$8 \times \tfrac{3}{4} = \$\tfrac{8}{4} \times 3 = \$6$; and $\$6 \div \$\tfrac{3}{4} = \$8$, as before.

(3) When both divisor and dividend are fractions. What quantity of cloth at $\$\tfrac{3}{20}$ a yard can be bought for $\$\tfrac{4}{5}$? The quantity is given in $\$\tfrac{4}{5} \div \$\tfrac{3}{20}$, where again the quantities must be expressed in terms of a common unit of measure: there results $\$\tfrac{16}{20} \div \$\tfrac{3}{20} = 16 \div 3 = 5\tfrac{1}{3}$, which is the number of yards.

If $5\tfrac{1}{3}$ yards of calico cost $\$\tfrac{4}{5}$, what is the price per yard? We have $\$\tfrac{4}{5} \div \tfrac{16}{3}$—that is, one third of 16 times some quantity = $\$\tfrac{4}{5}$; 16 times the quantity = $\$\tfrac{4}{5} \times 3$; the quantity is $\$\tfrac{4}{5} \times 3 \times \tfrac{1}{16} = \$\tfrac{3}{20}$.

The Inverted Divisor.—It is obvious in all these cases that practically the divisor has been inverted and then treated as a factor with the dividend to get the quotient. It must be clear, too, that this is simply *reducing the quantities to be compared to the same unit of measure.* When $12 is to be divided by $\$\tfrac{4}{5}$—i. e., when their ratio is to be found—they *must* be expressed in the same unit of measure. The divisor is measured off in *fifths* of a dollar; the dividend, then, must be expressed in *fifths* of a dollar—that is, it becomes 5×12, or **60**. The question is now changed to one of common division: $\$12 \div \tfrac{4}{5} = \tfrac{60}{5} \div \tfrac{4}{5} = 60 \div 4 = 15$. Similarly, in $\$\tfrac{4}{13} \div \$\tfrac{4}{5} = \tfrac{60}{65} = \tfrac{16}{13}$, the divisor is expressed in fifths of a dollar; the dividend $12\frac{2}{13}$ must be expressed in fifths; how is this done? By multiplying $12\frac{2}{13}$ by 5, which gives the number of *fifths* in $12\frac{2}{13}$, namely, $\frac{60}{13}$; for if $12 is 60 fifths, $\frac{1}{13}$ of $12 must be $\frac{60}{13}$ fifths. The unit of measure is now the same, and we have $\frac{60}{13}$ (fifths) ÷ 4 (fifths) = $1\frac{5}{13}$. "Inverting the divisor," then, makes the problem one of ordinary division by *expressing the quantities in the same number measure.*

Though formal proofs of rules are in general too abstract to begin with, yet after the pupil has freely used and learned the nature of the processes involved in concrete examples, he will quite readily comprehend the more abstract proof, and even the general demonstration. Take a few instances:

1. To prove that the product of two fractions has for its numerator the product of the numerators of the given fractions, and for its denominator the product of their denominators:

(1) Prove $\frac{3}{8} \times \frac{5}{9} = \frac{15}{72}$

$\frac{3}{8} \times 5 = \frac{15}{8}$; but this product is 9 times too great, and therefore the required product is $\frac{1}{9}$ of $\frac{15}{8} = \frac{15}{72}$.

(2) $\frac{3}{8} \times 8 = 3$; for $\frac{3}{8} \times 8 = 3 \times \frac{1}{8} \times 8$; and

$\frac{5}{9} \times 9 = 5$; multiply these equals;

∴ $\frac{3}{8} \times 8 \times \frac{5}{9} \times 9 = 3 \times 5$; divide by 8×9;

∴ $\frac{3}{8} \times \frac{5}{9} = \frac{3 \times 5}{8 \times 9}$; i. e., the product of numerators, etc.

(3) Generally, let $\frac{a}{b}$ and $\frac{c}{d}$ be any two fractions.

$\frac{a}{b} \times b = a$; (because, from the nature of number,

$\frac{1}{b} \times b = 1$); similarly,

$\frac{c}{d} \times d = c$; multiply these equals;

$\therefore \frac{a}{b} \times \frac{c}{d} \times bd = a \times c$; divide these equals by bd;

$\therefore \frac{a}{b} \times \frac{c}{d} = \frac{a \times c}{b \times d}$; that is, the product of the numerators of the given fractions is the numerator of the required fraction, and the product of their denominators its denominator.

2. To prove the rule for division of fractions, "invert the divisor and proceed as in multiplication."

(1) $\frac{3}{8} \div \frac{5}{9}$; divide $\frac{3}{8}$ by 5 and there results $\frac{3}{40}$; but it is required to divide not by 5 but by $\frac{1}{9}$ of 5; the required quotient must therefore be 9 times $\frac{3}{40}$—that is, $\frac{27}{40}$, which is $\frac{3 \times 9}{8 \times 5}$.

(2) $\frac{3}{8} \times 8 = 3$; multiply these equals by 9;

$\therefore \frac{3}{8} \times 8 \times 9 = 3 \times 9.$ \hfill (1)

Similarly, $\frac{5}{9} \times 9 = 5$; multiply these equals by 8;

$\therefore \frac{5}{9} \times 9 \times 8 = 5 \times 8.$ \hfill (2)

Divide (1) by (2):

$\therefore \frac{3}{8} \div \frac{5}{9} = \frac{3 \times 9}{5 \times 8}$.

Similarly, a general proof may be given, as in multiplication.

CHAPTER XIV.

DECIMALS.

As already indicated in Chapter X, decimals may be regarded as a natural and legitimate extension of the notation with which the pupils are already familiar. Taking this view of decimals as a basis for teaching the subject, we shall see how easily and naturally all the ordinary processes are established, and, further, how this mode of treatment recalls and confirms all that was said in building up the simple rules.

Notation and Numeration.—Consider the number 111: the first 1, starting at the right, denotes one unit; the second, *one* ten, or *ten* units; the third, *one* hundred, or *ten* tens, or *one hundred* units. The third 1 is equivalent to one hundred times the first 1, and to ten times the second 1; the second 1 is equivalent to ten times the first 1, and to one tenth of the third 1; the first 1 is equivalent to one tenth of the second 1, and to one hundredth of the third 1. Let us now rewrite the number already taken, place a point after the first 1 to indicate that that 1 is to be regarded as the unit, and then place after the point three 1's, so that we have

$$111 \cdot 111.$$

We may ask what each of these 1's should mean, if the same relation is to hold among successive digits that we have supposed hitherto to hold. The 1 after the point, standing next to the 1 which the point tells us is to be looked upon as a unit, would naturally mean one tenth of that 1—that is, one tenth of a unit, or, as we shall say, one tenth. The next 1, passing to the right, standing two places to the right of the unit, is one hundredth of the unit, or one hundredth; it is one tenth of the preceding 1—that is, one tenth of one tenth. Similarly, the next 1 would signify one thousandth, and would equal one hundredth of the one tenth or one tenth of the one hundredth. Thus the number above written may be read as follows: One hundred, one ten, one unit, one tenth, one hundredth, and one thousandth. But just as in ordinary numbers it is convenient, for the purpose of reading, to combine the elements into groups, here also it will be well to adopt a similar method. The 1 to the extreme right is 1 thousandth; the next 1 is, from its position, equivalent to 10 thousandths; and the next 1 is 100 thousandths; so that to the right of the point we have 111 thousandths. The whole number may now be read, one hundred and eleven, *and* one hundred and eleven thousandths. Very little practice will suffice to acquaint the pupil with the extended notation and numeration. A few questions, such as the following, will prove useful:

(1) Read 539·7423, and show that the reading properly expresses the number.

(2) Explain how it is that the insertion of a zero between the point and the 5 in the decimal ·5 changes the value of the decimal, but that the addition of zeros to the right of the 5 does not change the value.

(3) Name the decimal consisting of three digits which lies nearest in value to ·573245.

These will serve to bring out in a new relation some of the essential features of the decimal system, and throw light on some facts that at an earlier stage in the pupil's progress were necessarily somewhat dimly seen.

I. Simple Rules.

Multiplication.—When once the notation is understood, addition and subtraction of decimals can offer no difficulties, and we pass them by to consider multiplication. In this connection the most striking application is the multiplication by 10, 100, etc. The pupil will be asked to compare 7 and ·7, ·3 and ·03, ·009 and ·0009, and he will see at once that the first number in each case is, in virtue of the position of the point, 10 times the second number. Next, when asked to compare the numbers 37 and 3·7, he will see that the 3 in the first number is 10 times as great as the 3 in the second, that the 7 in the first number is 10 times as great as the 7 in the second, and that therefore the first number is 10 times as great as the second number. He has thus been led to discover that by moving the points one place to the right we get a number 10 times as great as the original number. Similarly, a corresponding conclusion may be reached for multiplication by 100, 1000, etc., and the conclusions in each case should be arrived at and stated by the pupil. It will at once follow that to divide a number by 10, 100, 1000, etc., we have only to move the point one place, two places, three places, etc., to the left. We pass next to the multiplication by any integral number.

$$5\cdot37$$
$$\underline{3}$$
$$16\cdot11$$

$$4\cdot42$$
$$\underline{57}$$
$$30\cdot94$$
$$\underline{221\cdot0}$$
$$251\cdot94$$

The multiplication in each of the foregoing cases is based on the same considerations as the multiplication of integers by integers. Thus, in the second case, 7 times 2 hundredths are 14 hundredths—that is, 1 tenth and 4 hundredths, and the 4 must be in the hundredths place; 7 times 4 tenths are 28 tenths, which with the former 1 tenth make up 29 tenths or 2 units and 9 tenths, and the 9 must be in the tenths place; thus, the 4 and the 9 will be properly placed if the point is introduced before the 9, etc.; next, multiplying by 5, we must write the results one place to the left, for reasons explained in an earlier chapter. The pupil will now understand multiplication by an integer, and is ready to proceed with multiplication by a decimal.

$$3\cdot1\,2$$
$$\underline{2\,3}$$
$$9\cdot3\,6$$
$$\underline{6\,2\cdot4}$$
$$7\,1\cdot7\,6$$

$$3\cdot1\,2$$
$$\underline{2\cdot3}$$
$$7\cdot1\,7\,6$$

He will be asked to multiply some number, say 3·12, by some number, say 23; the result is 71·76. If, then, we propose to multiply 3·12 by 2·3, it will be seen that this differs from the former only in that the multiplier is 10 times as small; the product then will be 10 times as small, and may at once be written down 7·176. A further example or two, in which a different number of decimal places are taken, will suffice to show that to multiply two decimals we proceed as in the multiplication of integers, and mark off in the resulting product as many places as there are in both multiplier and multiplicand.

Division.—To teach division, it is well to begin with the division by an integer, as this will connect the process with what is already known. Consider the following examples:

```
      (1)                    (2)
   7)2 1(3               7)2·1(·3
     2 1                   2·1
     ───                   ───

      (3)                    (4)
5)·0 0 1 5(·0 0 0 3    2 3)1 4 5·8 1(5·4 7
  ·0 0 1 5                 1 3 5
  ──────                   ─────
                           1 0·8
                             9 2
                           ─────
                           1·6 1
                           1·6 1
                           ─────
```

The pupil who can explain the first division can at once explain the second, the third, and the fourth; and he will see how to divide whenever the divisor is a whole number. Then he may be asked to explain why, in the following divisions, we have the same quotient:

```
   3)15              12)60
     ──                 ──
      5                  5
```

He will be led to recognise a principle that he already knows, namely, that the multiplication of divisor and dividend by the same number does not change the quotient. He may then be asked to state a quotient equivalent to $35 \div \cdot 7$, but having for divisor a whole number. An answer to be expected is, $350 \div 7$; at any rate, he can be led to this result, and, as this quotient is seen to be 50, he can conclude that $35 \div \cdot 7 = 50$. An examination of a few more examples will show how always to proceed.

It would be well to have the student, in his earlier practice, write out a full statement of what he does. Suppose he is required to find the quotient $1\cdot3754 \div \cdot 23$; his solution should stand somewhat as follows:

The quotient of $1\cdot3754$ by $\cdot 23$ is the same as (multiplying each number by 100) that of $137\cdot54$ by 23.

$$
\begin{array}{r}
2\,3)1\,3\,7\cdot5\,4(5\cdot98 \\
1\,1\,5 \\ \hline
2\,2\cdot5 \\
2\,0\cdot7 \\ \hline
1\cdot8\,4 \\
1\cdot8\,4 \\ \hline
\end{array}
$$

$\therefore 1\cdot3754 \div \cdot 23 = 5\cdot98$.

The Relation of Decimals to Vulgar Fractions.— The simple rules being understood, we may now consider the conversion of decimals to the ordinary fractional form, and the conversion of ordinary fractions into the decimal form.

From the definition, $\cdot 273 = \frac{2}{10} + \frac{7}{100} + \frac{3}{1000} = \frac{273}{1000}$, and the student sees at once how to write a decimal in the form of a fraction.

We may next ask the pupil to divide 1 by 2, as an exercise in the division of decimals.

$$
\begin{array}{r}
2)1\cdot0(\cdot5 \\
1\cdot0 \\ \hline
\end{array}
$$

Similarly,

$$4\overline{)3{\cdot}00}({\cdot}75$$
$$\underline{2{\cdot}8}$$
$$\overline{20}$$

$$8\overline{)7{\cdot}00}({\cdot}875$$
$$\underline{6{\cdot}4}$$
$$\overline{60}$$
$$\underline{56}$$
$$\overline{40}$$

But the quotients $1 \div 2$, $3 \div 4$, $7 \div 8$ have up to this point been taken as equivalent to $\frac{1}{2}$, $\frac{3}{4}$, $\frac{7}{8}$.

$\therefore \frac{1}{2} = {\cdot}5$, $\frac{3}{4} = {\cdot}75$, $\frac{7}{8} = {\cdot}875$.

As an exercise these results might be verified thus:

$${\cdot}75 = \tfrac{75}{100} = \tfrac{3}{4}.$$

A method has now been found for converting an ordinary fraction into a decimal; at the same time another method has been suggested in the verification above made; for we see that $\frac{3}{4} = \frac{3}{2 \times 2} = \frac{3 \times 5 \times 5}{2 \times 5 \times 2 \times 5} = \frac{75}{100} = {\cdot}75$. The latter method is very valuable from the point of view of theory, and the pupil should work several examples in this way.

We shall next consider the example $\frac{2}{3}$.

$$3\overline{)2{\cdot}00}({\cdot}66$$
$$\underline{1{\cdot}8}$$
$$\overline{20}$$
$$\underline{18}$$
$$\overline{2}$$

$\therefore \frac{2}{3} = {\cdot}6666 \ldots$, the 6's being repeated without end. This fact is expressed thus: $\frac{2}{3} = {\cdot}\dot{6}$, and $\frac{2}{3}$ is said to give rise to a recurring decimal. Let us now seek to convert $\frac{2}{3}$ to a decimal by the other method. According to it, we multiply the denominator by some number which will change it into 10, 100, 1000, etc.—that is, into some power of 10. Now, any such power is made by multiplying 10 by itself some number of times; but 10 itself is made by multiplying 2 and 5; therefore every power of 10 is made up wholly of the factors 2 and 5, and in equal number. We can not, then, multiply 3 by any number that will make it into a power of ten—that is, we can not convert $\frac{2}{3}$ into an ordinary decimal with a finite number of digits. We have thus a complete view of the case. The following are examples of recurring decimals:

$$\frac{1}{7} = \cdot\dot{1}4285\dot{7} \qquad \frac{1}{9} = \cdot\dot{1} \qquad \frac{1}{11} = \cdot\dot{0}\dot{9}.$$

Take next $\frac{1}{6}$:

$$6)1\cdot000(\cdot 166$$
$$\underline{6}$$
$$40$$
$$\underline{36}$$
$$40$$
$$\underline{36}$$
$$4$$

$\therefore \frac{1}{6} = \cdot 16666 \ldots$, the 6's recurring, and this is written $\frac{1}{6} = \cdot 1\dot{6}$.

Here the first figure of the decimal does not recur, and $\frac{1}{6}$ is said to give rise to a *mixed* recurring decimal, those formerly met with being called *pure* recurring decimals. Similarly,

$$\frac{7}{12} = \cdot 58\dot{3} \qquad \frac{1}{18} = \cdot 0\dot{5}.$$

If we try to apply the second method to these examples, we get—

$$\frac{1}{6} = \frac{1}{2 \times 3} = \frac{5}{2 \times 5 \times 3} = \frac{1}{10} \text{ of } \frac{5}{3}$$

$$\frac{7}{12} = \frac{7}{2 \times 2 \times 3} = \frac{7 \times 5 \times 5}{5 \times 2 \times 5 \times 2 \times 3} = \frac{1}{100} \text{ of } \frac{175}{3}$$

$$\frac{1}{18} = \frac{1}{2 \times 3 \times 3} = \frac{5}{5 \times 2 \times 3 \times 3} = \frac{1}{10} \text{ of } \frac{5}{9}.$$

DECIMALS.

The examples given lead up to the following propositions, for the truth of which it will be easy to state the general argument:

Proposition I.—A fraction whose denominator contains only the factors 2 or 5 leads to a decimal, the number of whose digits is the same as the number of times the factor 2 or the factor 5 is contained in the denominator, according as the former factor or the latter occurs the greater number of times.

Proposition II.—A fraction whose denominator contains neither the factor 2 nor the factor 5 leads to a pure recurring decimal.

Proposition III.—A fraction whose denominator contains in addition to the factors 2 and 5 a factor prime to these factors, leads to a mixed recurring decimal, the number of digits that are before the period being the same as the number of times the factor 2 or the factor 5 is contained in the denominator, according as the former factor or the latter occurs the greater number of times.

Questions similar to the following afford a valuable exercise on this part of the work:

(1) In the case of fractions, such as $\frac{1}{7}$, $\frac{1}{29}$, etc., leading to pure recurring decimals, what limit is there to the number of figures in the period?

(2) $\frac{1}{7} = \cdot\dot{1}4285\dot{7}$: explain why any other fraction with denominator 7 will lead to a recurring decimal with a period consisting of the same digits following one another in the same circular order.

It will now be in place to consider the converse process of changing recurring decimals into their equivalent vulgar fractions. A difference of opinion exists as to the best mode of dealing with these decimals. The method here presented is for many reasons thought to be the best:

$$.\dot{3} = .3333\ldots \text{ (the 3's repeated without end);}$$
$$\therefore .\dot{3} \times 10 = 3.3333\ldots \text{ (the 3's repeated without end);}$$
$$\text{and } .\dot{3} = .3333\ldots \text{ (the 3's repeated without end);}$$

$\therefore$ taking $.\dot{3}$ from 10 times $.\dot{3}$ we have

$$.\dot{3} \times 9 = 3$$
$$\therefore .\dot{3} = \tfrac{3}{9} = \tfrac{1}{3}$$

which may be tested by converting $\tfrac{1}{3}$ into decimal form. Similarly,

$$\therefore .\dot{1}\dot{7} \times 100 = 17.17171717\ldots \left.\begin{matrix}\\\\\end{matrix}\right\} \text{(17 repeated}$$
$$.\dot{1}\dot{7} = .17171717\ldots \quad\text{without end);}$$

$\therefore$ subtracting,

$$.\dot{1}\dot{7} \times 99 = 17; \therefore .\dot{1}\dot{7} = \tfrac{17}{99}$$

which in its turn may be tested.

Next, to find the fractional equivalent of $.2\dot{7}\dot{9}$

$$.2\dot{7}\dot{9} = .2797979\ldots \left.\begin{matrix}\\\\\\\end{matrix}\right\} \text{(79 repeated}$$
$$\therefore .2\dot{7}\dot{9} \times 1000 = 279.797979\ldots \quad\text{without end);}$$
$$\text{and } .2\dot{7}\dot{9} \times 10 = 2.797979\ldots$$

$\therefore$ subtracting,

$$.2\dot{7}\dot{9} \times 990 = 279 - 2 = 277$$
$$\therefore .2\dot{7}\dot{9} = \frac{277}{990}$$

The pupil, after working a few examples, will be in a position to formulate a rule for writing down the fraction which is equal to any given recurring decimal.

This treatment has the advantage of furnishing the pupil a direct and definite method of procedure. Against

DECIMALS. 271

it is urged the fact that there is made an appeal to infinite series, and to the notion of a limit. In reply to this, it may be said that in the natural way of finding the decimal which is equal to a fraction we come upon this infinite series—indeed, we can not avoid it. Further, the notion of a limit—the term need not be given to the class—is, after all, not a difficult one; it may be difficult to establish in the case of any given series that it has a limit, and more difficult still to find that limit; but the difficulty is not in the idea of limit.

CONTRACTED METHODS.

There are few processes that lend themselves so readily to teaching as these contracted methods. It should be explained to the class that, very frequently in physics and in actual business life, operations have to be performed with decimals, often with many digits in the decimal part, when there are required in the results only a few decimal places. For example, if the answer to a commercial problem were \$79.5917235, it would be taken to be \$79.59—that is to say, we should take the result as far as two places. In the same way, if a required length were 59·37542156 centimetres, and the graduated rule made it impossible to consider anything beyond thousandths of a centimetre, we should take the result as 59·375. Accordingly, if we know in advance that a result of four places, say, is all that is required, we may find it possible to avoid unnecessary calculations.

Contracted Multiplication. — Example: Multiply 4·2578532 by 7 correct to four places.

The pupil might be expected to place the 7, the multiplier, beneath the 8 of the multiplicand. If he does not regard what is "carried" from the multiplication of the 5, the complete multiplication might be performed and the results compared:

4·2578532	4·2578532
7	7
29·8049	29·8049724

This one example would serve to show that we must take into account what may be carried into the result by the multiplication of the digit to the right of the place directly considered. After the pupil has worked a few examples similar to this, he might then be asked to multiply 4·2578532 by 40 correct to four places. It will be seen that multiplying 5 (in the fifth place) by 1 ten will give 5 in the fourth place, so that if we place 4—which is 4 tens—beneath the 5 and commence the multiplication there, we shall have a result reaching to four places. Working also in full, the need for taking account of the "carried" number will appear.

4·2578532	4·2578532
4	40
170·3141	170·3141280

So, too, the multiplication of the same number by **300** correct to four places, will be seen to be

4·2578532	4·2578532
3	300
1277·3559	1277·3559600

Next take the multiplication of 4·2578532 by ·5, cor-

DECIMALS. 273

rect to four places. The 7 thousandths of the multiplicand, if multiplied by 1 tenth, will give 7 in the fourth place, so that we may begin the multiplication with the 7.

$$
\begin{array}{r} 4{\cdot}2578532 \\ 5 \\ \hline 2{\cdot}1289 \end{array}
\qquad
\begin{array}{r} 4{\cdot}2578532 \\ {\cdot}5 \\ \hline 2{\cdot}12892660 \end{array}
$$

If, now, we have to multiply 4·2578532 by 347·5 correct to four places, all that is necessary is to bring together the four results arrived at already:

$$
\begin{array}{r}
4{\cdot}2578532 \\
5743 \\
\hline
1277{\cdot}3559 \\
170{\cdot}3141 \\
29{\cdot}8049 \\
2{\cdot}1289 \\
\hline
1479{\cdot}6038
\end{array}
$$

But now let us compare this with the complete multiplication:

$$
\begin{array}{r}
4{\cdot}2578532 \\
347{\cdot}5 \\
\hline
21289\,2660 \\
298049\,724 \\
1703141|28 \\
12773559|6 \\
\hline
1479{\cdot}6039|8700
\end{array}
$$

It will be seen that not *all* that is necessary has been taken into account, inasmuch as there is a difference in the results—a difference of 1 in the fourth place. The reason for this can at once be seen: the addition (in the complete multiplication) of the digits in the fifth place contributes 1 more than we expected to be carried. To guard against this error, a certain precaution is taken which may easily be explained. Suppose we are multiplying by 9, and we have to multiply 7 by it to get the number to be carried; this gives 6 to carry, the 3 belonging to the next place and therefore being overlooked. Suppose, in the same multiplication, we have to multiply by 6, and we have to multiply 8 by it to get the number to carry; this gives 4 to carry, the 8 belonging to the next place and therefore being overlooked. But already we have overlooked 3 in that place, making in all **11**, which would make a difference of 1 in the place to which our result is to be correct. On this account, when we get such numbers as 27, 56, 48, 49—that is, with their second digit greater than 5—in the multiplication which is to give the number to be carried, we consider them as giving to carry 3, 6, 5, 5; whereas such numbers as 32, 42, 54 give 3, 4, 5. One has to use one's judgment about the number to be carried from such numbers as 25, 35, 45. One is, however, liable to overestimate or underestimate, and to secure accuracy to the fourth place, say, it is generally best to multiply to the fifth place.

Contracted Division.—Suppose we have to divide 5·8792314 by 3·421384 correctly to 3 decimal places. Immediately, or by multiplying both numbers so as to make the divisor a whole number, we see that the first significant figure of the quotient is in the units place, so that we have to find a quotient consisting of four digits, three of which are to the right of the decimal point. First let the ordinary division be performed:

$$3{\cdot}421384)5{\cdot}879\,2314(1{\cdot}718$$
$$\underline{3{\cdot}421\,384}$$
$$2{\cdot}457\,8474$$
$$\underline{2{\cdot}394\,9688}$$
$$62\,8760$$
$$\underline{34\,21384}$$
$$28\,664760$$
$$27\,371072$$

It is plain that to get the last figure of the quotient we needed only the first two figures of the third remainder; so that, if a vertical line were drawn immediately to the right of these two figures, we retain to the left quite enough to determine all the figures of the quotient, unless, indeed, the subtractions, etc., that affect the column of figures to the right of this line may affect the numbers to the left of the line. It will be noticed that to the left of the line (in this case) in the dividend is equal to the number of figures in our result. Let us now try to construct that part of the work which lies to the left of the line. The first step will stand thus:

$$3{\cdot}42138\overline{4})5{\cdot}879(1$$
$$\underline{3{\cdot}421}$$
$$2{\cdot}458$$

and this means that we consider only the first four figures of the divisor (there being nothing to carry from the multiplication of the fifth). But the remainder differs by 1 in the last place, which is due to the "carrying" in the subtraction to the right of the line in the original division; however, let this difference be overlooked for the present. The next division, in the figures to the left of the line, is concerned with only three figures of the divisor, there being nothing to carry from the multiplication of 1 by 7; we shall now have the work thus:

$$3{\cdot}42\cancel{1})5{\cdot}879(1{\cdot}7$$
$$\underline{3{\cdot}421}$$
$$2{\cdot}458$$
$$\underline{2{\cdot}394}$$
$$64$$

Here the 1 of the divisor was marked out after the first division. It will be noticed that the remainder is here 64, while the corresponding number in the complete division is 62. Referring to the original work we see that, so far as the figures to the left of the line are concerned, we have to do only with the first two figures of the divisor. We may therefore strike out the third figure, and our work will stand thus:

$$3{\cdot}4\cancel{2}\cancel{1})5{\cdot}879(1{\cdot}71$$
$$\underline{3{\cdot}421}$$
$$2{\cdot}458$$
$$\underline{2{\cdot}394}$$
$$64$$
$$\underline{34}$$
$$30$$

Here the remainder is 30, while the corresponding remainder in the complete work is 28. Now strike out, for the same reason as before given, the 4 of the divisor; we are then in doubt whether the last figure should be 9 or 8, as, taking 9, we see that $9 \times 3 = 27$, which, with the 3 to carry from the multiplication of 4 by 9, makes 30, but we might suspect the 9 to be too great. We see now that if we wish to have four figures, we should start with a divisor of four figures. For the same reasons as given in the case of multiplication, we should also adopt a similar rule for carrying; and, further, if we wish our answer to consist of four figures, we are more likely to be strictly correct to that place if we start with a divisor of five figures, which means that we retain an additional column of figures of the original division. The work would then stand thus:

$$3{\cdot}42138)5{\cdot}8792(1{\cdot}718$$
$$\underline{3{\cdot}4214}$$
$$2{\cdot}4578$$
$$\underline{2{\cdot}3949}$$
$$629$$
$$\underline{342}$$
$$287$$
$$\underline{273}$$

We have thus to regard the following:

(1) Find the number of figures that are to be in the answer.

(2) Start the division with a divisor consisting of a number of digits one more than that number, these digits to be the first digits of the given divisor in order.

(3) After each subtraction, instead of placing a figure (from the dividend) to the right of the remainder, cut off one figure from the right of the divisor.

(4) In multiplying, have regard always to what may be carried from the neglected digit to the right, regard-

ing such a number as 48 as giving 5 to carry, such a number as **32** as giving 3.

Manifestly, as in the case of multiplication, there is need of practice to give one confidence, and to educate one's judgment in the matter of deciding what number should be carried in such doubtful cases as may arise.

CHAPTER XV.

PERCENTAGE AND ITS APPLICATIONS.

Percentage.—In some text-books on arithmetic percentage is treated as if it were a special process involving certain distinctive principles and therefore entitled to rank as a separate department. In these books, accordingly, percentage has its definitions, its "cases," and its rules and formulas. This elaborate treatment seems to be a mistake on both the theoretical and the practical side: on the theoretical side, because it asserts or assumes a new phase in the development of number; on the practical side, because it substitutes a system of mechanical rules for the intelligent application of a few simple principles with which the student is perfectly familiar. In the growth of number as measurement percentage presents nothing new. It has to do with the ideas and processes of ratio with which fractions are more or less explicitly concerned, and its problems afford excellent practice for enlarging and defining these ideas, and securing greater facility in using them. But the mere fact that, in this new rule with its cases and its rules, a quantity is measured off into a hundred parts instead of into any other possible number of parts, appears to be no valid reason for constituting percentage a new process marking a new phase in the evolution of number. It is no doubt correct enough to say that "percentage is a process of computing by hundredths"; but is such a process to be broadly distinguished as a mental operation, from a process of computing by eighths, or tenths, or twentieths, or fiftieths? If the difference between fractions and percentage is not a difference in logical or psychological process, but chiefly a difference in handling number symbols, is it worth while to invest the subject with an air of mystery, and invent, for the edification of the pupil, from six to nine "cases" with their corresponding rules and formulas?

The real facts regarding percentage indicate clearly enough that, to say the least, there is no need for this formal treatment and the complexity to which it gives rise.

(1) The phrase *per cent*, a shortened form of the Latin per centum, is equivalent to the English word hundredths; and a rate per cent is, then, simply a number expressing so many hundredths of a quantity. Thus, 1 per cent, 2 per cent, 3 per cent, 4 per cent . . . n per cent means 1, 2, 3, 4 . . . n of the hundred equal parts into which a given quantity may be divided, just as $\frac{1}{50}, \frac{2}{50}, \frac{3}{50} \ldots \frac{n}{50}$ means 1, 2, 3 . . . n of the fifty parts into which a quantity may be divided; or, in general, as $\frac{1}{n}, \frac{2}{n}, \frac{3}{n}, \frac{4}{n} \ldots$ represents 1, 2, 3, 4 . . . of the n parts into which a quantity may be measured off.

(2) All problems in percentage involve, then, simply the principles discussed in fractions, and may be solved by direct application of these principles. Indeed, for the mental work with *which every arithmetical "rule" should be introduced,* and by which its study should be constantly accompanied, the easiest and most effective

PERCENTAGE AND ITS APPLICATIONS. 281

treatment is, in general, by means of the simplest forms of fractions relatively to the quantities involved, whether *hundredths*, or *nineteenths*, or *twentieths*, or *n*ths.

(3) All the so-called cases in percentage are therefore but direct applications of fractions as expressing definite measurement, or at least may be easily solved by such applications. Take the following brief illustrations of the principal "cases":

1. *To find any given per cent of a quantity.*

(*a*) A dealer purchased a quantity of goods at $\frac{3}{5}$ of their wholesale price, which was $325; what did he pay for the goods?

(*b*) A dealer purchased a quantity of goods at 60 per cent of their wholesale price, which was $325; how much did he pay for the goods?

Here (*a*) is a problem in fractions and (*b*) one in percentage; in the one case a certain quantity is expressed as 3-fifths of another; in the other it is expressed as 60-hundredths of it.

2. *To find what per cent one quantity is of another.*

(*a*) What part (fraction) of $325 is $195?

(*b*) What per cent of $325 is $195?

In a certain sense question (*a*) may be said to be indefinite—i. e., any one of an unlimited number of equivalent fractions may be taken as a correct answer. Thus, the answer is $\frac{195}{325} = \frac{39}{65} = \frac{3}{5} (= \frac{780}{1300} = \frac{60}{100}$, etc.). But if the question were, how many 325ths of $325 in $195? How many 65ths of it? How many fifths of it?—the respective answers to each of these are the first three of these fractions, and they are all found by exactly the same reasoning. For example: 1. $\frac{1}{325}$ of $325 is $1; this is contained 195 times in $195; therefore, $195 is $\frac{195}{325}$ of $325. 2. $\frac{1}{65}$ of $325 is $5; this is contained 39 times in $195; therefore, $195 is $\frac{39}{65}$ of $325. 3. $\frac{1}{5}$ of $325 is $65; this is contained 3 times in $195; therefore, $195 is $\frac{3}{5}$ of $325. Similarly for other equivalent fractions which answer corresponding questions.

In question (*b*) we are, strictly speaking, limited to one answer, but it is found in exactly the same way; may, in fact, be obtained from any of the unlimited series of fractions that answer question (*a*).

The question really is, how many hundredths of $325 are there in $195? We reason as before: $\frac{1}{100}$ of $325 is $3¼ (or $3.25); this is contained 60 times in $195; therefore, $195 is $\frac{60}{100}$ of $325. The solutions of these questions might, of course, have been varied by *first* multiplying $195 by 1, 65, 5, and 100 respectively; thus, in question (*a*) the comparison is to be made between $195 and the hundredth of $325—i. e., how often is $$\frac{325}{100}$$ contained in $195, where (see Division of Fractions) the quantities must be expressed in the same unit of measure, and the division is 19500 (hundredths of $1) ÷ 325 (hundredths of $). In general, the most direct way is to find any convenient fraction expressing the ratio of the quantities, and then change this to an equivalent fraction having 100 for denominator.

3. *To find the number of which a certain per cent is given.*

(*a*) A dealer bought goods for $195, which was $\frac{3}{5}$ of cost; find the cost.

(*b*) A dealer bought goods for $195, which was 60 per cent of cost; find the cost.

In (*a*) the cost is measured off in 5 equal parts, and

PERCENTAGE AND ITS APPLICATIONS.   283

3 of them are given : 3 of them = $195, 1 of them = $65, 5 of them (the whole) = $65 × 5 = $325. In (*b*) the cost is conceived of as measured off in 100 equal parts, and 60 of them are given : 60 of them = $195, 1 of them = $195 ÷ 60, 100 of them (the whole) = $195 ÷ 60 × 100 = $325. Here, as in the last case, in accordance with the principle connecting factors and divisors, we might have multiplied by the respective factors before dividing by the respective divisors—e. g., 5 times $\frac{1}{3}$ of a quantity = $\frac{1}{3}$ of 5 times the quantity—that is, $65 ÷ 3 × 5 = $65 × 5 ÷ 3.

Introductory Lesson.—Different teachers will use different devices in applying in percentage the simple principles of fractions. The following points are merely suggested :

1. It will hardly be necessary, at this stage of the pupil's development, to use concrete illustrations. It will certainly not be necessary if the pupil has been taught arithmetic according to the psychology of the subject. Begin the teaching of arithmetic with the use of things, but do not continue and end with things. So long as pupils have to use objects, they are apt to attend to the mere practical processes at the expense of the higher mental processes through which alone number concept can arise. The infantile stage of dependence on objects is only a stage ; it is not to be a permanent resting place ; the method of crawling on all-fours may seriously arrest development.

2. The first aim will be to get the pupil to identify per cents with fractions. He already knows how and why a fraction may be changed to an equivalent fraction having any given numerator or denominator. (1)

Give, then, exercises expressing certain simple fractions in (exact number of) hundredths: $\frac{1}{2} = \frac{50}{100}$; $\frac{1}{4} = \frac{25}{100}$; $\frac{1}{5} = \frac{20}{100}$; $\frac{1}{10} = \frac{10}{100}$; $\frac{1}{20} = \frac{5}{100}$; $\frac{3}{4} = \frac{75}{100}$; $\frac{3}{5} = \frac{60}{100}$; $\frac{7}{10} = \frac{70}{100}$, etc. It will readily be seen that a large number of fractions can be changed into equivalent *single* fractions having 100 for a denominator; in other words, into fractions expressing an exact number of hundredths. (2) Then some exercises to show that any fraction may be expressed in hundredths:

$$\frac{1}{3} = \frac{33\frac{1}{3}}{100}; \quad \frac{1}{8} = \frac{12\frac{1}{2}}{100}; \quad \frac{7}{8} = \frac{87\frac{1}{2}}{100}; \quad \frac{19}{21} = \frac{1900}{2100} = \frac{90\frac{10}{21}}{100}, \text{ etc.}$$

The pupils already know that multiplication and division by ten and by a hundred are very easily performed; in other words, that a number of tenths or a number of hundredths is more easily found than any other fraction of that quantity; they will also see that the number of fractions that can be expressed as a whole number of hundredths is much larger than the number that can be expressed as a whole number of tenths; they will probably infer why the practice of measuring off a quantity in hundredths has been so generally adopted.

3. The different ways of writing hundredths will be recalled, and the symbol for the phrase *per cent* will be given; for example, 5 per cent has the symbol 5 per cent, and is expressed by $\frac{5}{100}$, 5 hundredths, and 0·5.

4. Easy mental problems (followed by written work) connecting fractions with percentage, and illustrating the different "cases" of percentage. What fractions are equivalent to the following: 1 per cent, 10 per cent, 25 per cent, 30 per cent, 60 per cent, 80 per cent, 90 per cent, etc.? What per cent of a quantity is $\frac{1}{2}$ of it, $\frac{1}{4}$ of

it, $\frac{1}{2}$ of it, $\frac{1}{3}$ of it, $\frac{1}{4}$ of it, $\frac{2}{3}$ of it, $\frac{3}{8}$ of it? Questions like these, together with practical problems in the same line, will serve to show the identity in principle between fractions and percentage. Percentage is but another name for fractions.

5. The pupils will be then prepared for more formal problems illustrating the general cases. These are not to be presented as *special* cases demanding special rules, definitions, and formulas. The thing is to avoid multiplying rules and hair-splitting definitions, and to give the pupil facility in the application of a few simple principles. It has been proved by actual experience that students who never heard of the nine cases of percentage, and the nine rules or formulas, have readily acquired the power to handle any problem in percentage except, perhaps, such as, on account of their complexity, are more properly exercises in algebraic analysis. The pupil should not be confined to any one mode of solution in working problems in percentage. He will sometimes use the purely fractional form, at others the so-called percentage form, and in still other cases a combination of both forms. He should be so instructed in the real nature of the principles and practised in their application as to be able to use all forms with equal facility, and almost instantly determine, in any given problem, which of the forms will lead to the most concise and elegant solution.

It may be well to utter a caution against the vague use of the phrase *per cent*, which too generally prevails. It is often used as if it possessed in itself a clear and definite meaning. It denotes simply a possible mode of measurement. Ten per cent, or one hundred per cent, has no more meaning than ten or one; all numbers signify *possible* measurements; they are empty of meaning till applied to measured *quantity*. It is not uncommon to find in published solutions of percentage problems *different* quantities used as defined by the same unit of measure because they are expressed in per cents. We have before us, for example, a solution in which the author takes it for granted that the difference between 110 per cent of one quantity and 90 per cent of a different quantity is 20 per cent. "Let 100 per cent equal the required quantity" is a very common presupposition in the solution of a percentage problem, and equally common to it to find the same 100 per cent "doing" duty for some other quantity which demands recognition in the same solution. So, in a recent English work of great pretensions, we have it posted, in all the emphasis of black letter—as a fundamental working principle—that "**100 per cent is 1.**" One hundred per cent of any quantity—like 2-halves of it, or 3-thirds of it, or 4-fourths of it . . . , or n-$n$ths of it—is indeed the quantity taken *once*, or one time. But this loose way of making "100 per cent equal to 1," or to any quantity, is due to a total misconception of the nature of number as measurement of quantity, and of the function of the fraction as stating explicitly the process of measurement. It seems as if both teachers and pupils were often hypnotized by this subtle *one hundred per cent*.

Some Applications of Percentage.

1. *Profit and Loss.*—We do not need either formal cases or formal rules, as "given the buying price and

PERCENTAGE AND ITS APPLICATIONS. 287

the selling price to find the gain or loss per cent." A few examples will serve to illustrate the different "cases."

(1) Bought sugar for 6 cents per pound and sold it for 8 cents per pound; find the gain *per cent*.

The question simply stated is: 2 cents gain on 6 cents cost means how much gain on 100 cents of cost; that is, $\frac{2}{6}$ = how many hundredths? Multiply both terms by $16\frac{2}{3}$ ($= \frac{100}{6}$). *Or*,

   6 cents outlay gains 2 cents;
∴ 1 cent " " $\frac{1}{3}$ cent;
∴ 100 cents " " $\frac{1}{3} \times 100 = 33\frac{1}{3}$ cents.

(2) Cloth was bought at 60 cents a yard, and sold to gain 25 per cent; find the selling price.

Take the cost price as unit of comparison: Selling price = 125 per cent of cost = $\frac{5}{4}$ cost = $\frac{5}{4}$ of 60 cents = 75 cents. *Or*,

  On 100 cents of cost gain is 25 cents.
  On 1 cent of cost gain is $\frac{25}{100}$ "

∴ On 60 cents of cost gain is $\frac{1}{4}$ cent $\times$ 60 = 15 cents. Hence 60 cents + 15 cents = 75 cents, the selling price.

(3) By selling cotton at 12 cents a yard there is a gain of 20 per cent; what was the cost price?

Take the cost price as unit of comparison: 20 per cent of cost is $\frac{1}{5}$ of cost; therefore, $1\frac{1}{5}$ cost = $\frac{6}{5}$ cost = 12 cents. Therefore, cost = 10 cents.

(4) By selling coffee at 30 cents a pound a grocer lost 25 per cent; what price would bring him a profit of 10 per cent?

Selling price = $\frac{3}{4}$ of cost = 30 cents; therefore, cost = 40 cents. New price = $\frac{11}{10}$ of cost = $\frac{11}{10}$ of 40 cents = 44 cents. Otherwise, the losing price, $\frac{3}{4}$ ($\frac{15}{20}$) of cost, must be increased to $\frac{11}{10}$ ($\frac{22}{20}$) of cost—that is, must be increased in the ratio $\frac{22}{15}$; therefore, $\frac{22}{15}$ of 30 cents = 44 cents, the price required.

(5) A merchant gains 30 per cent by selling goods at 39 cents a yard; at what selling price would he lose 40 per cent?

Gaining price is $\frac{13}{10}$ of cost. Losing price is $\frac{6}{10}$ of cost; therefore the latter is $\frac{6}{13}$ of the former = $\frac{6}{13}$ of 39 = 18 (cents).

2. *Stocks, Commission, etc.*—A few examples will show that there is no new principle in these rules.

(1) How much cash will be realized by selling out $4,000 stock, Government 5's, at $95\frac{1}{4}$?

$100 stock brings $95\frac{1}{4}$ cash; $4,000 stock brings $95\frac{1}{4} \times 40 = $3,810 cash.

(2) What amount will be realized by selling out $4,400 six-per-cent stock at $106\frac{3}{8}$, allowing brokerage $\frac{1}{8}$?

Every 100 of stock brings $(106\frac{3}{8} - \frac{1}{8}) = $106\frac{2}{8}$; therefore, $4,400 of stock brings $106\frac{1}{4} \times 44 = $4,675.

(3) What semi-annual income will be derived from investing $9,000 in bank stock selling at $120 and paying 4 per cent half yearly dividends?

$120 will buy $100 stock, which brings $4 income— that is, the income is $4 \div 120 = \frac{1}{30}$ of the investment = $\frac{1}{30}$ of $9,000 = $300.

(4) Which is the better investment, a stock paying 12 per cent at $140, or one paying 9 per cent at $120? What income from investing $1,400 in each?

In the first investment $140 brings $12 income; therefore, $1 brings $$\frac{12}{140} = $\frac{3}{35}$.

In the second investment $120 brings $9; there-

PERCENTAGE AND ITS APPLICATIONS.   289

fore, $1 brings $\frac{9}{120} = \frac{3}{40}$; $\frac{3}{35}$ is greater than $\frac{3}{40}$; therefore, the first is the better investment.

Income from the first $= \frac{3}{35}$ of $1,400 = $120;
Income from the second $= \frac{3}{40}$ of $1,400 = $105.

(5) A commission merchant is instructed to invest $945 in certain goods, deducting his commission of $2\frac{1}{2}$ per cent on the price paid for the goods; find the agent's commission.

Since the agent receives $2\frac{1}{2}$ for every $100 he invests, $102\frac{1}{2}$ must be sent for every $100 that is to be invested in goods; that is, for every $102\frac{1}{2}$ sent, the agent receives $2\frac{1}{2}$; therefore, he receives $2\frac{1}{2} \div 102\frac{1}{2} = \frac{1}{41}$ of the whole amount sent; therefore, amount of commission $= $945 \div 41$.

(6) For how much must a house worth $3,900 be insured at $2\frac{1}{2}$ per cent, so that the owner, in case of loss, may recover both the value of the house and the premium paid?

Since the premium is $2\frac{1}{2}$ per cent of the amount insured, the property must be 100 per cent — $2\frac{1}{2}$ per cent $= 97\frac{1}{2}$ per cent $= \frac{39}{40}$ of the amount insured; therefore, $\frac{39}{40}$ of this amount $= $3,900, and the amount is $4,000.

(7) What amount must a town be assessed so that after allowing the collector 2 per cent the net amount realized may be $24,500?

The collector gets 2 per cent $= \frac{1}{50}$ of total levy; therefore, town gets $\frac{49}{50}$ of total levy; therefore, $\frac{49}{50}$ of total levy $= $24,500, and, therefore, total levy $= $25,000.

INTEREST.

The pupil, having learned the meaning and the use of the term *per cent*, should find very little difficulty in the subject of interest. However in the problems of interest and kindred commercial work pupils frequently fail; but if the cause of the failure is examined into, it will nearly always be found to be, not so much an inability to meet the mathematics of the problems, as a want of accurate knowledge of the terms used, and of acquaintance with the business forms and operations involved. On this account, in taking up the applications of arithmetic to commercial work, the teacher should be at great pains to ensure that every pupil understands well, and sees clearly through, all such forms and operations.

Simple Interest.—In accordance with what has been said, it is necessary first to explain to the class how men, when loaning money, require a certain payment for the use of the money, and how the amount to be paid for this use—that is to say, the interest—depends on the time, twice, thrice, etc., the time (implying, as it does, twice, thrice, etc., the use), requiring twice, thrice, etc., the interest. The unit of time is generally taken as one year, and the *rate* for the year is given as a per cent. Accordingly, if we say that a man loans money for a year at 5 per cent per annum, we mean that at the end of one year he would receive as interest $\frac{5}{100}$ of the money loaned; if the money were loaned for half a year the interest would be $\frac{1}{2}$ of $\frac{5}{100}$ of the sum loaned, and if for fifty-three days, it would be $\frac{53}{365}$ of $\frac{5}{100}$ of the sum loaned. The pupil is now prepared to do any problem of calculation of simple interest, and after being trained in the formal working and stating of such problems—that is, after realizing the problem to the full—should be trained in making rapid calculations after the methods of men in business.

He should next be led to see the relations among the interest, the sum loaned (the principal), and the sum called the amount. Suppose the sum loaned to be $100, the time to be six months, and the rate 6 per cent per annum. Take the line A B to represent six months, A the beginning of the time, and B the end of the time.

$$\begin{array}{ll} & \text{\$3 interest.} \\ \text{Principal: \$100} & \text{\$100 principal.} \\ \overline{\quad A \quad\quad\quad\quad\quad\quad\quad B \quad} \end{array}$$

At the end of the time the sum $103 has to be paid to the loaner—that is, the $100 has to be restored, and $3 paid as interest. The sum, $103, is called the amount. It is plain then that

(1) The interest $= \frac{3}{100}$ of the principal;
$\qquad\qquad\quad = \frac{3}{103}$ of the amount.
(2) The principal $= \frac{100}{3}$ of the interest;
$\qquad\qquad\quad = \frac{100}{103}$ of the amount.
(3) The amount $= \frac{103}{3}$ of the interest;
$\qquad\qquad\quad = \frac{103}{100}$ of the principal.

The use of a line to represent time will assist the pupil greatly, and after examining a few examples similar to the foregoing, he will *know* all the relations among principal, interest, and amount, and will see how to write them down when the rate and the time are given. When these relations are understood, the whole subject of interest is understood, the only care required on the part of the teacher consisting in making a careful gradation of problems.

One of the most striking applications of interest is to problems relating to the so-called *true discount*—a term which should fall into disuse. There is but one discount, the discount of actual business life; it is an application of percentage, and on account of its being calculated in the same way as interest it is erroneously spoken of as interest, and a confusion arises in the mind of the pupil.

Accordingly, the problem, Find what sum would pay now a debt of $150 due at the end of six months, the *rate of interest* being 6 per cent per annum, is a definite problem in interest. To solve it we have recourse to the line illustration given above. It is plain that if one had $100 now, and put it out at *interest* at the rate given, it would come back at the end of the time as $103. Thus, $100 now is the equivalent of $103 at the end of six months—that is, the sum now, equivalent to a certain sum due at the end of six months, is $\frac{100}{103}$ of that sum. Therefore, in the case in question, the sum is $\frac{100}{103}$ of $150. It is true that there is here an allowance off, a discount, so to speak, but until the pupil understands the whole question of interest and discount the term should not be used in this connection. We shall suppose, then, that the student has mastered simple interest, and shall turn to compound interest.

Compound Interest.—The teacher should explain that the value of money—as the pupil has seen—depends, in some measure, on where it is placed in time; men in business always suppose interest to be paid when it is due, or if an agreement is made that its payment be deferred, they regard this interest in its turn a source of interest. An example worked out in detail will help the pupil to see just what is done. Suppose a sum of $1,000 loaned for three years, at 5 per cent per annum,

PERCENTAGE AND ITS APPLICATIONS. 293

interest to be paid at the end of the three years, and the interest at the end of each year to become a source of interest for the ensuing year or years:

```
    $1 0 0 0    Principal.
           5
    ─────────
    $5 0.0 0    First year's interest.
    $1 0 0 0
    $1 0 5 0    Sum bearing interest for the 2d year.
           5
    ─────────
    $5 2.5 0    Second year's interest.
    $1 0 5 0.
    $1 1 0 2.5 0   Sum bearing interest for the 3d year.
             5
    ─────────
    $5 5.1 2 5 0   Third year's interest.
    1 1 0 2.5 0
    ─────────
    $1 1 5 7.6 2 5 0   Amount to be paid at end of time.
    1 0 0 0.0 0        Original principal.
    ─────────
    $1 5 7.6 2 5       Amount of interest.
```

The pupil will work several such examples, and will find not a little pleasure in determining just how much interest has been paid as interest on interest. He is then ready to make a more general study of compound interest.

Suppose a sum loaned at compound interest for three years at 5 per cent per annum. What is the interest on any sum for one year at 5 per cent? Plainly $\frac{5}{100}$ of the sum. What is the amount? $\frac{105}{100}$ of the sum. What, then, is the amount of any sum for one year? $\frac{105}{100}$ of *that* sum. What sum bears interest for the second year? $\frac{105}{100}$ of the original sum. What will the amount of this be? $\frac{105}{100}$ of itself, and therefore $\frac{105}{100}$ of $\frac{105}{100}$ of the original sum. Accordingly, the amount of the sum for two years is $(\frac{105}{100})^2$ of the original sum. What for three years? Plainly $\frac{105}{100}$ of $(\frac{105}{100})^2$ of original sum, and therefore $(\frac{105}{100})^3$ of original sum. This is found to be $\frac{1157625}{1000000}$ of original sum. How much more have we than the original sum? $\frac{157625}{1000000}$ of original sum; therefore, interest = $\frac{157625}{1000000}$ of sum = $\frac{157625}{1157625}$ of amount, etc.

The pupil should be told that in all transactions involving a time longer than one year (or it may be by agreement six months or three months) compound interest is alone employed where the interest is thought of as all being paid at the end of the time. From what has been said he will know at once how to solve the following problem of interest: Find what sum paid now will discharge a debt of $1,000, due at the end of three years, the *rate of interest* being 6 per cent. He should acquire a facility in thus transferring money from one time to another.

Annuities.—A few words may be said on the subject of annuities. If A gives B $100 to keep for all time, and the rate of interest be 6 per cent, B would be undertaking an equivalent if he would agree (for himself and his heirs) to pay to A (and his heirs) $6 at the end of each year, for all time. This $6 supposed paid at the end of each year is called an annuity; as it runs for all time, it is called a perpetual annuity, and is said to begin now, though the first payment is made at the end of the first year. The $100 is very properly called its cash value, and the relation of the $100 to the annuity of $6 is plainly that of principal to interest. Thus, it

PERCENTAGE AND ITS APPLICATIONS. 295

will be easy to find the cash value of any given perpetual annuity, or to find the perpetual annuity that could be purchased with a given sum. To illustrate this we should need a line extending beyond all limits:

$\mid\!\!—\!\!\mid\!\!—\!\!\mid\!\!—\!\!\mid\!\!—\!\!\mid\!\!—\!\!\mid\!\!—\!\!\mid\!\!—\!\!\mid\!\!—\!\!\mid\!\!—\!\!\mid\!\!—\!\!\mid\!\!>\cdots$
$100 $6 $6 $6 $6 $6 $6 $6 $6

(The divisions of the line represent each one year.)

Next we may suppose an annuity to begin at the end of, say, three years, so that the first payment would be made at the end of the fourth year. Taking the annuity to be $6 and the rate 6 per cent per annum, we see that the value of this annuity at the beginning of the fourth year (represented by the point in the illustration below) is $100.

 $100 $6 $6 $6 $6
$\mid\!\!—\!\!\mid\!\!—\!\!\mid\!\!—\!\!\mid\!\!—\!\!\mid\!\!—\!\!\mid\!\!—\!\!\mid\!\!—\!\!\mid\!\!>\cdots$
A   B   C   D   E   F   G   H

But that $100 is placed at the end of three years from now, and is therefore equivalent to $(\frac{100}{106})^3$ of $100 now. We have thus the cash value of an annuity *deferred three* years.

When the pupil knows how to deal with the two cases discussed he can easily be led to find the cash value of an annuity beginning now and running for a definite number of years. When asked to compare the two perpetual annuities represented below, he will see that the first exceeds the second by three payments—$6 at the end of the first year, $6 at the end of the second year, and $6 at the end of the third year, and these constitute an annuity for three years beginning now.

```
 $6  $6  $6  $6  $6  $6  $6  $6  $6
|___|___|___|___|___|___|___|___|___> ...
             $6  $6  $6  $6  $6  $6
 _____|___|___|___|___|___|___> ...
```

But the cash value of the first annuity is $100, and the cash value of the second is $(\frac{100}{108})^3$ of $100.

∴ The cash value of an annuity of $6 beginning now, running for three years, is $100 − $(\frac{100}{108})^3$ of $100 or $\{1 - (\frac{100}{108})^3\}$ of $100.

It will be easy to obtain a general formula, and also to find the value of a deferred annuity running for a definite number of years.

CHAPTER XVI.

EVOLUTION.

Square Root.—The product of 3 and 3 is 9; of 5 and 5 is 25. The measures of squares whose sides measure 3 and 5 are 9 and 25. We say that 9 is the square of 3, and that 25 is the square of 5; 3 is the square root of 9, and 5 the square root of 25. The square of 3 is written 3^2, the square root of 3 is expressed thus: $\sqrt{3}$. The pupil can write at once the table of squares:

$$1^2 = 1$$
$$2^2 = 4$$
$$3^2 = 9$$
$$4^2 = 16$$
$$5^2 = 25 \quad [10^2 = 100]$$
$$6^2 = 36$$
$$7^2 = 49$$
$$8^2 = 64$$
$$9^2 = 81$$

He will note that the square of any number expressed by one digit is a number expressed by one digit or by two digits, while the lowest number expressed by two digits—viz., 10—has for its square **100**, a number expressed by three digits.

It is plain that the square of any number expressed by two digits has for its square a number expressed by three digits or by four digits. Also the square root of a number* expressed by three digits is a number expressed by two digits, and the tens digit is known from the first digit on the left; for example, 625 (if it has an exact square root), lying as it does between 400 and 900, will have for square root a number lying between 20 and 30—that is, the tens figure of the root will be 2. Similarly, if a number is expressed by four digits its square root is expressed by two digits, and the tens digit of the root can be determined from the first two digits (to the left) of the number; thus the square root of 2709—a number lying between 2500 and 3600—will have 5 for a tens digit, and this is determined by the **27** of the number 2709.

Write next the table of squares:

$$10^2 = 100$$
$$20^2 = 400$$
$$30^2 = 900$$
$$40^2 = 1600$$
$$50^2 = 2500 \qquad [100^2 = 10000]$$
$$60^2 = 3600$$
$$70^2 = 4900$$
$$80^2 = 6400$$
$$90^2 = 8100$$

Now take 13 and square it: the result is 169. We wish to arrive at a method of recovering 13 from 169.

---

* In general, when we speak of the square root of a number, we suppose that it has an exact square root.

EVOLUTION.

To do this we shall examine how the 169 is formed from the 13:

$$\begin{array}{r} 13 \\ 13 \\ \hline 9 \\ 30 \\ 30 \\ 100 \\ \hline 169 \end{array}$$

Thus 13, which is made up of two parts, 10 and 3, has for its square a number 169, which is seen to be made up of 100, the square of 10; 9, the square of 3; and twice the product of 10 and 3. This is familiar to the pupil who has worked algebra, and may be illustrated geometrically. The whole square is measured by 13^2, and its parts make up $10^2 + 2 \times (10 \times 3) + 3^2$.

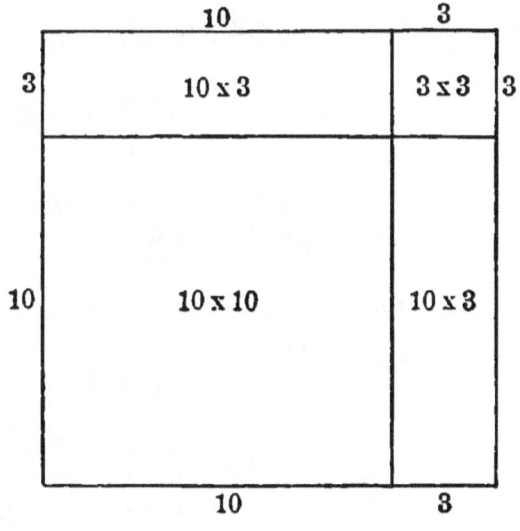

Now, to recover 13 from 169: we see that its hundreds digit 1, showing that the number lies between 100 and 400, gives the tens digit of the root, so that we know one of the parts of the root, viz., 10. The square of this part is 100, and the rest of the given number, 69, must be 2 times 10, multiplied by the other part, together with the square of the other part.

$$\begin{array}{r}169 | 10 \\ \underline{100} \\ 69\end{array}$$

Accordingly, if we multiply 10 by 2 and divide 69 by this product we get a clue to the other part. Dividing 69 by 2 × 10, or 20, we see that the quotient is a little greater than 3; if, then, after taking 3 times 20 from 69 there is left the square of 3, we have the root. Plainly this is the case:

$$\begin{array}{r} 169 | 10+3 \\ \underline{100} \\ 20 \quad \overline{|\ 69\ } \\ \underline{|\ 60\ } \\ 9 = 3^2 \end{array}$$

Now, this work might be written somewhat more neatly, thus:

$$\begin{array}{r} 169 | 10+3 \\ \underline{100} \\ \left.\begin{array}{r}20 \\ +3\end{array}\right\} \quad \overline{|\ 69\ } \\ |\ 69\ \end{array}$$

It may be further simplified by leaving out unnecessary zeros, thus:

$$\begin{array}{r} 169 | 13 \\ \underline{1} \\ 23 \quad \overline{|\ 69\ } \\ |\ 69\ \end{array}$$

The pupil is now in a position to find the square root of all numbers expressed by three or four digits. It would be well, before considering the square root of larger numbers, to examine for the square root of such numbers as 1·69, 27·09. The pupil will see at once that the square root of the former number lies between 1 and 2, that of the latter between 5 and 6, and can easily be led to complete the process of extracting the roots, finding as results 1·3 and 5·3. He will thus discover for himself that the problem is not different from the one already solved.

We are now ready to examine for the square root of larger numbers. Write first the following table:

$$100^2 = 10000$$
$$200^2 = 40000$$
$$300^2 = 90000$$
$$400^2 = 160000$$
$$500^2 = 250000 \quad [1000^2 = 1000000]$$
$$600^2 = 360000$$
$$700^2 = 490000$$
$$800^2 = 640000$$
$$900^2 = 810000$$

A study of the table will lead to the conclusion that the square roots of numbers expressed by 5 or 6 digits are numbers expressed by 3 digits, and that, if expressed by 5 digits, the first (to the left) digit of the root is determined by the first (to the left) figure of the number, and if expressed by 6 digits, by the first two digits of the number. Thus, the square root of 16900 will have 1 as the hundreds digit, while that of 270900 will have 5 as the hundreds digit. Next, by multiplication, we find that

$$230^2 = 52900$$
and $$240^2 = 57600$$

Consequently the square root of (say) 54756 must lie between 230 and 240—that is, must have 23 as its first two digits. The first three digits of the number 54756 are sufficient to show that this must be the case. Now, suppose we seek the square root of 54756.

Plainly, the first part of the root is 200:

$$\begin{array}{r|l} 54756 & 200 \\ 40000 & \\ \hline 14756 & \end{array}$$

Now, had we been seeking the square root of 52900, which is 230—that is, consists of two parts, 200 and 30—we should have worked thus:

$$\left.\begin{array}{r} 200 \times 2 = 400 \\ 30 \end{array}\right\} \begin{array}{|r} 52900 | 200 + 30 \\ 40000 \\ \hline 12900 \\ 12900 \end{array}$$

Similarly for 57600:

$$\left.\begin{array}{r} 200 \times 2 = 400 \\ + \; 40 \end{array}\right\} \begin{array}{|r} 57600 | 200 + 40 \\ 40000 \\ \hline 17600 \\ 17600 \end{array}$$

Then plainly we see how, in finding the square root of 54756, to determine the second figure:

$$\left.\begin{array}{r} 200 \times 2 = 400 \\ + \; 30 \end{array}\right\} \begin{array}{|r} 54756 | 200 + 30 \\ 40000 \\ \hline 14756 \\ 12900 \\ \hline 1856 \end{array}$$

We have yet to find the units digit of the root. But at this point we may say that the root consists of two parts, one 230, and the other to be found, and may proceed as in the earlier case:

$$
\begin{array}{r|l}
& 54756\underline{|200+30+4} \\
& 40000 \\ \hline
400 & 14756 \\
+\,30 & \\ \hline
430 & 12900 \\
230\times 2 = 460 & 1856 \\
+\,4 & \\ \hline
464 & 1856 \\
\end{array}
$$

The work may now be shortened:

$$
\begin{array}{r|l}
& 54756\underline{|234} \\
& 4 \\ \hline
43 & 147 \\
& 129 \\ \hline
464 & 1856 \\
& 1856 \\
\end{array}
$$

After the pupil has been exercised in extracting the roots of numbers expressed by 5 or 6 digits, he will find no difficulty in determining the square roots of such numbers as 547·56, 5·4756, ·054756. The extension to numbers expressed by a higher number of digits will be easy, and the need for marking off into periods of two, starting from the decimal point, as well as its full significance, will have been realized by the pupil.

Up to this point we have spoken of numbers whose

square root can be extracted; it will be next in order to deal with the approximations to square roots—for example, the square root of 2, 5, etc.; but as this involves nothing essentially new it will not be here discussed.

We shall conclude this part of the work by calling attention to the extraction of the square root of a fraction. Since

$$\frac{2}{3} \times \frac{2}{3} = \frac{4}{9},$$

the square root of $\quad \dfrac{4}{9} = \dfrac{\sqrt{4}}{\sqrt{9}} = \dfrac{2}{3}.$

In the case of fractions whose denominators are numbers whose roots can not be exactly determined, we should proceed as follows:

Take, for example, $\quad \sqrt{\dfrac{2}{3}}:$

$$\sqrt{\frac{2}{3}} = \sqrt{\frac{2 \times 3}{3 \times 3}} = \frac{\sqrt{6}}{3},$$

an artifice the value of which is apparent.

Cube Root.—The method of teaching square root has been presented in such detail that very few words will suffice on the subject of cube root.

From examples such as $3^3 = 27$ the meaning of cube and cube root will be brought out, and use may be made of the geometrical illustration of the cube. The pupil should commit to memory the following table:

$1^3 = 1$
$2^3 = 8$
$3^3 = 27$
$4^3 = 64$
$5^3 = 125$ $[10^3 = 1000]$
$6^3 = 216$
$7^3 = 343$
$8^3 = 512$
$9^3 = 729$

All numbers expressed by 1, 2, or 3 digits have for cube roots numbers expressed by 1 digit.

Next we have the following table:

$10^3 = 1000$
$20^3 = 8000$
$30^3 = 27000$
$40^3 = 64000$
$50^3 = 125000$ $[100^3 = 1000000]$
$60^3 = 216000$
$70^3 = 343000$
$80^3 = 512000$
$90^3 = 729000$

Thus all numbers expressed by 4, 5, or 6 digits have for cube roots numbers expressed by 2 digits; further, the first digit of the root in such case is determined by the first one (to the left), the first two, or the first three digits of the number, according as it is expressed by 4, 5, or 6 digits. Thus the cube roots of the numbers 2744, 39304, 357911, will in every case be numbers expressible by two digits. The tens digits will be determined by the 2, the 39, the 357 of the numbers to be 1, 3, 7 respectively.

To find the cube root of a number we shall see how the cube of a number is formed. The identity of the following two ways of multiplying 14 by 14, and the product by 14, will at once be seen:

$$
\begin{array}{rr}
10 + 4 & 14 \\
10 + 4 & 14 \\
\hline
(4 \times 10) + 4^2 & 56 \\
10^2 + (4 \times 10) & 14 \\
\hline
10^2 + 2(4 \times 10) + 4^2 & 196 \\
10 + 4 & 14 \\
\hline
(4 \times 10^2) + 2(4^2 \times 10) + 4^3 & 784 \\
10^3 + 2(4 \times 10^2) + (4^2 \times 10) & 196 \\
\hline
10^3 + 3(4 \times 10^2) + 3(4^2 \times 10) + 4^3 & 2744
\end{array}
$$

We see, then, that the cube of such a number as 14—that is, a number regarded as being made up of two parts, here 10 and 4—is the cube of one part increased by three times the product of the square of that part and the other part, and three times the product of the first part and the square of the second, and the cube of the second part.

We wish now to recover from 2744 its cube root. Plainly, the tens digit of the root is 1—that is, the first part of the root is ten; take from 2744 the cube of 10:

$$
\begin{array}{r}
2744 | \underline{10} \\
1000 \\
\hline
1744
\end{array}
$$

The remainder is made up of three parts:

(1) The product of 3 times the square of 10, and the other part of the root.

(2) The product of 3 times 10, and the square of the other part of the root.

EVOLUTION. 307

(3) The cube of the other part of the root.

Then, if we divide 1744 by 3 times the square of 10, we shall have a clue to the other part of the root. Dividing, we may take 5 as the other part:

$$3 \times 10^2 = 300 \quad \begin{array}{r|l} 2744 & 10+5 \\ 1000 & \\ \hline 1744 & \\ 1500 & \\ \hline 244 & \end{array}$$

Taking away $3 \times 10^2 \times 5$, we have as remainder 244, which should be made up of parts (2) and (3), mentioned above. We find, however, that it is not large enough. We have taken a second part too large, and therefore take a smaller part, say 4:

$$3 \times 10^2 = 300 \quad \begin{array}{r|l} 2744 & 10+4 \\ 1000 & \\ \hline 1744 & \\ 1200 & \\ \hline 544 & \end{array}$$

Here the remainder is 544, which $= 3 \times 10 \times 4^2 + 4^3$, and we conclude that the root is 14.

We see also that the $1744 = 4(3 \times 6^2 + 4 \times 3 \times 10 + 4^2)$. The work might then be shown thus:

$$\left.\begin{array}{r} 3 \times 10^2 = 300 \\ 4 \times 3 \times 10 = 120 \\ 4^2 = 16 \\ \hline 436 \end{array}\right\} \quad \begin{array}{r|l} 2744 & 10+4 \\ 1000 & \\ \hline 1744 & \\ \\ 1744 & \end{array}$$

It may be further shortened, thus:

$$
\begin{array}{r|l}
2744 & \underline{14} \\
1 & \\
\hline
& 1744 \\
\end{array}
$$

$$
\begin{aligned}
1^2 \times 300 &= 300 \\
1 \times 4 \times 30 &= 120 \\
4^2 &= \underline{16} \\
&436
\end{aligned}
\quad \Big| \ 1744
$$

The further development of the method will follow lines similar to those followed in square root. We shall take space only to indicate how, in the case of finding a cube root consisting of several figures, a certain saving of work may be secured:

$$
\begin{array}{r|l}
814\ 780\ 504 & \underline{934} \\
729 & \\
\hline
85\ 780 & \\
\end{array}
$$

$$
\begin{aligned}
90^2 \times 3 &= 24300 \\
3 \times 90 \times 3 &= 810 \\
3^2 &= 9 \\
\hline
3 \times 90^2 + 3 \times (3 \times 90) + 3^2 &= 25119 \\
&9 \\
\hline
&2594700 \\
&11160 \\
&16 \\
\hline
&2605876
\end{aligned}
\quad
\begin{array}{|l}
75\ 357 \\
\hline
10\ 423\ 504 \\
\\
\\
\\
\hline
10\ 423\ 504
\end{array}
$$

When we reach the point where we wish to determine the third figure, we have to find three times the square of 93—that is,

$$3\,(90^2 + 2 \times 90 \times 3 + 3^2)$$

Now, as indicated above, 810 is $3 \times 90 \times 3$, 9 is 3^2, 25119 is $3 \times 90^2 + 3 \times 90 \times 3 + 3^2$, so that, if to the sum of 810, 9, 25119 we add 9, which is the square of

3, we shall have found three times the square of 93. If to the resulting number we affix two zeros, we shall have three hundred times the square of 93.

This artifice may be employed when at each successive stage we need three hundred times the square of the part already found.

We shall conclude this chapter with the remark that the fourth root of a number is to be found by extracting the square root of its square root, and the sixth root of a number by extracting the square root of its cube root.

THE END.

D. APPLETON & CO.'S PUBLICATIONS.

NEW VOLUMES IN THE INTERNATIONAL EDUCATION SERIES.

FRIEDRICH FROEBEL'S PEDAGOGICS OF THE KINDERGARTEN; or, *His Ideas concerning the Play and Playthings of the Child.* Translated by JOSEPHINE JARVIS. International Education Series. 12mo. Cloth, $1.50.

This book holds the keynote of the "New Education," and will assist many in a correct comprehension of the true principles underlying the practical outcome of Froebel's thought. Although extant for nearly fifty years, his ideas are still in need of elucidation, and the average kindergartner and primary-school teacher grasps but a superficial meaning of the methods suggested.

SYSTEMATIC SCIENCE TEACHING. A Manual of Inductive Elementary Work for all Instructors in Graded and Ungraded Schools, the Kindergarten, and the Home. By EDWARD GARDNIER HOWE. 12mo. Cloth, $1.50.

A thoroughly practical and reliable guide to elementary instruction in science has long been a desideratum, and this work, embodying the results of fourteen years of actual classroom tests, will satisfactorily meet such a demand. The volume gives a general outline of work for the first three years.

THE EDUCATION OF THE GREEK PEOPLE, and its Influence on Civilization. By THOMAS DAVIDSON. 12mo. Cloth, $1.50.

"This work is not intended for scholars or specialists, but for that large body of teachers throughout the country who are trying to do their duty, but are suffering from that want of enthusiasm which necessarily comes from being unable clearly to see the end and purpose of their labors, or to invest any end with sublime import. I have sought to show them that the end of their work is the redemption of humanity, an essential part of that process by which it is being gradually elevated to moral freedom, and to suggest to them the direction in which they ought to turn their chief efforts. If I can make even a few of them feel the consecration that comes from single-minded devotion to a great end, I shall hold that this book has accomplished its purpose."
—*Author's Preface.*

THE EVOLUTION OF THE MASSACHUSETTS PUBLIC-SCHOOL SYSTEM. A Historical Sketch in Six Lectures. By GEORGE H. MARTIN, A. M., Supervisor of Public Schools, Boston, Massachusetts. 12mo. Cloth, $1.50.

The public discussion that arose from Professor Martin's papers upon the topic treated in this work will make the complete collection of the essays of much interest to a large circle of readers. In the present volume the author aims to show the evolutionary character of the public-school history of the State, and to point out the lines along which the development has run and the relation throughout to the social environment, and incidentally to illustrate the slow and irregular way by which the people under popular governments work out their own social and intellectual progress.

New York: D. APPLETON & CO., 72 Fifth Avenue.

D. APPLETON & CO.'S PUBLICATIONS.

Recent Volumes of the International Scientific Series.

A HISTORY OF CRUSTACEA. By Rev. THOMAS R. R. STEBBING, M. A., author of "The Challenger Amphipoda," etc. With numerous Illustrations. 12mo. Cloth, $2.00.

"Mr. Stebbing's account of 'Recent Malacostraca' (soft-shelled animals) is practically complete, and is based upon the solid foundations of science. The astonishing development of knowledge in this branch of natural history is due to the extension of marine research, the perfecting of the microscope, and the general diffusion of information regarding what has been ascertained concerning the origin of species. . . . This volume is fully illustrated, and contains useful references to important authorities. It is an able and meritorious survey of recent crustacea."—*Philadelphia Ledger.*

HANDBOOK OF GREEK AND LATIN PALÆOGRAPHY. By EDWARD MAUNDE THOMPSON, D. C. L., Principal Librarian of the British Museum. With numerous Illustrations. 12mo. Cloth, $2.00.

"Mr. Thompson, as principal librarian of the British Museum, has of course had very exceptional advantages for preparing his book. . . . Probably all teachers of the classics, as well as specialists in palæography, will find something of value in this systematic treatise upon a rather unusual and difficult study."—*Review of Reviews.*

"Covering as this volume does such a vast period of time, from the beginning of the alphabet and the ways of writing down to the seventeenth century, the wonder is how, within three hundred and thirty-three pages, so much that is of practical usefulness has been brought together."—*New York Times.*

MAN AND THE GLACIAL PERIOD. By G. FREDERICK WRIGHT, D. D., LL. D., author of "The Ice Age in North America," "Logic of Christian Evidences," etc. With numerous Illustrations. 12mo. Cloth, $1.75.

"The author is himself an independent student and thinker, whose competence and authority are undisputed."—*New York Sun.*

"It may be described in a word as the best summary of scientific conclusions concerning the question of man's antiquity as affected by his known relations to geological time."—*Philadelphia Press.*

RACE AND LANGUAGE. By ANDRÉ LEFÈVRE, Professor in the Anthropological School, Paris. 12mo. Cloth, $1.50.

"A most scholarly exposition of the evolution of language, and a comprehensive account of the Indo-European group of tongues."—*Boston Advertiser.*

"A welcome contribution to the study of the obscure and complicated subject with which it deals."—*San Francisco Chronicle.*

"One of the few scientific works which promise to become popular, both with those who read for instruction and those who read for recreation."—*Philadelphia Item.*

New York: D. APPLETON & CO., 72 Fifth Avenue.

D. APPLETON & CO.'S PUBLICATIONS.

THE ANTHROPOLOGICAL SERIES.

"Will be hailed with delight by scholars and scientific specialists, and it will be gladly received by others who aspire after the useful knowledge it will impart."—*New York Home Journal.*

NOW READY.

WOMAN'S SHARE IN PRIMITIVE CULTURE. By OTIS TUFTON MASON, A. M., Curator of the Department of Ethnology in the United States National Museum. With numerous Illustrations. 12mo. Cloth, $1.75.

"A most interesting *résumé* of the revelations which science has made concerning the habits of human beings in primitive times, and especially as to the place, the duties, and the customs of women."—*Philadelphia Inquirer.*

"A highly entertaining and instructive book. . . . Prof. Mason's bright, graceful style must do much to awaken a lively interest in a study that has heretofore received such scant attention."—*Baltimore American.*

"The special charm of Mr. Mason's book is that his studies are based mainly upon actually existing types, rather than upon mere tradition."—*Philadelphia Times.*

THE PYGMIES. By A. DE QUATREFAGES, late Professor of Anthropology at the Museum of Natural History, Paris. With numerous Illustrations. 12mo. Cloth, $1.75.

"Probably no one was better equipped to illustrate the general subject than Quatrefages. While constantly occupied upon the anatomical and osseous phases of his subject, he was none the less well acquainted with what literature and history had to say concerning the pygmies. . . . This book ought to be in every divinity school in which man as well as God is studied, and from which missionaries go out to convert the human being of reality and not the man of rhetoric and text-books."—*Boston Literary World.*

"It is fortunate that American students of anthropology are able to enjoy as luminous a translation of this notable monograph as that which Prof. Starr now submits to the public."—*Philadelphia Press.*

"It is regarded by scholars entitled to offer an opinion as one of the half-dozen most important works of an anthropologist whose ethnographic publications numbered nearly one hundred."—*Chicago Evening Post.*

THE BEGINNINGS OF WRITING. By W. J. HOFFMAN, M. D. With numerous Illustrations. 12mo. Cloth, $1.75.

This interesting book gives a most attractive account of the rude methods employed by primitive man for recording his deeds. The earliest writing consists of pictographs which were traced on stone, wood, bone, skins, and various paperlike substances. Dr. Hoffman shows how the several classes of symbols used in these records are to be interpreted, and traces the growth of conventional signs up to syllabaries and alphabets— the two classes of signs employed by modern peoples.

IN PREPARATION.

THE SOUTH SEA ISLANDERS. By Dr. SCHMELTZ
THE ZUÑI. By FRANK HAMILTON CUSHING.
THE AZTECS. By Mrs. ZELIA NUTTALL.

New York: D. APPLETON & CO., 72 Fifth Avenue.

D. APPLETON & CO.'S PUBLICATIONS.

*L*IFE OF SIR RICHARD OWEN. By Rev. RICHARD OWEN. With an Introduction by T. H. HUXLEY. 2 vols. 12mo. Cloth, $7.50.

"The value of these memoirs is that they disclose with great minuteness the daily labors and occupations of one of the foremost men of science of England."—*Boston Herald.*

"A noteworthy contribution to biographical literature.'—*Philadelphia Press.*

*D*EAN BUCKLAND. The Life and Correspondence of WILLIAM BUCKLAND, D. D., F. R. S., sometime Dean of Westminster, twice President of the Theological Society, and first President of the British Association. By his Daughter, Mrs. GORDON. With Portraits and Illustrations. 8vo. Buckram, $3.50.

"Next to Charles Darwin, Dean Buckland is certainly the most interesting personality in the field of natural science that the present century has produced."—*London Daily News.*

"A very readable book, for it gives an excellent account, without any padding or unnecessary detail, of a most original man."—*Westminster Gazette.*

*S*CHOOLS AND MASTERS OF SCULPTURE. By A. G. RADCLIFFE, author of "Schools and Masters of Painting." With 35 full-page Illustrations. 12mo. Cloth, $3.00.

"The art lover will find in Miss Radcliffe's work a book of fascinating interest, and a thoroughly painstaking and valuable addition to the stock of knowledge which he may possess on the history of the noble art of sculpture."—*Philadelphia Item.*

"It would be difficult to name another work that would be so valuable to the general reader on the same subject as this book."—*San Francisco Bulletin.*

"The work is free of all needless technicalities, and will be of intense interest to every intelligent reader, while of inestimable value to the student of art."—*Boston Home Journal.*

BY THE SAME AUTHOR.

*S*CHOOLS AND MASTERS OF PAINTING. With numerous Illustrations and an Appendix on the Principal Galleries of Europe. New edition, fully revised, and in part rewritten. 12mo. Cloth, $3.00; half calf, $5.00.

"The volume is one of great practical utility, and may be used to advantage as an artistic guide book by persons visiting the collections of Italy, France, and Germany for the first time."—*New York Tribune.*

New York: D. APPLETON & CO., 72 Fifth Avenue.

D. APPLETON & CO.'S PUBLICATIONS.

THE LIBRARY OF USEFUL STORIES.

Messrs. D. Appleton & Co. have the pleasure of announcing a series of little books dealing with various branches of knowledge, and treating each subject in clear, concise language, as free as possible from technical words and phrases.

The volumes will be the work of writers of authority in their various spheres, who will not sacrifice accuracy to mere picturesque treatment or irrelevant fancies, but endeavor to present the leading facts of science, history, etc., in an interesting form, and with that strict regard to the latest results of investigation which is necessary to give value to the series.

Each book will be complete in itself. Illustrations will be introduced whenever needed for the just comprehension of the subject treated, and every care will be bestowed on the qualities of paper, printing, and binding.

The price will be forty cents per volume.

NOW READY.

THE STORY OF THE STARS. By G. F. CHAMBERS, F. R. A. S., author of "Handbook of Descriptive and Practical Astronomy," etc. With 24 Illustrations.

"Such books as these will do more to extend the knowledge of natural science among the people than any number of more elaborate treatises."—*Cincinnati Tribune.*

"An astonishing amount of information is compacted in this little volume."—*Philadelphia Press.*

THE STORY OF PRIMITIVE MAN. By EDWARD CLODD, author of "The Story of Creation," etc.

"This volume presents the results of the latest investigations into the early history of the human race. The value of an up-to-date summary like this is especially marked in view of the interest of the subject. It is written in clear, concise language, as free as possible from technical words and phrases. The author is a recognized authority, and his lucid text is accompanied by a large number of attractive illustrations."

THE STORY OF THE PLANTS. By GRANT ALLEN, author of "Flowers and their Pedigrees," etc.

IN PREPARATION.

THE STORY OF THE EARTH. By H. G. SEELEY, F. R. S., Professor of Geography in King's College, London. With Illustrations.

THE STORY OF THE SOLAR SYSTEM. By G. F. CHAMBERS, F. R. A. S.

New York: D. APPLETON & CO., 72 Fifth Avenue.

D. APPLETON & CO.'S PUBLICATIONS.

"No library of military literature that has appeared in recent years has been so instructive to readers of all kinds as the Great Commanders Series, which is edited by General James Grant Wilson."—*New York Mail and Express.*

GREAT COMMANDERS. A Series of Brief Biographies of Illustrious Americans. Edited by General JAMES GRANT WILSON. 12mo, cloth, gilt top, $1.50 per volume.

This series forms one of the most notable collections of books that has been published for many years. The success it has met with since the first volume was issued, and the widespread attention it has attracted, indicate that it has satisfactorily fulfilled its purpose, viz., to provide in a popular form and moderate compass the records of the lives of men who have been conspicuously eminent in the great conflicts that established American independence and maintained our national integrity and unity. Each biography has been written by an author especially well qualified for the task, and the result is not only a series of fascinating stories of the lives and deeds of great men, but a rich mine of valuable information for the student of American history and biography.

The volumes of this series thus far issued, all of which have received the highest commendation from authoritative journals, are:

ADMIRAL FARRAGUT. By Captain A. T. MAHAN, U. S. N.
GENERAL TAYLOR. By General O. O. HOWARD, U. S. A.
GENERAL JACKSON. By JAMES PARTON.
GENERAL GREENE. By Captain FRANCIS V. GREENE, U. S. A.
GENERAL J. E. JOHNSTON. By ROBERT M. HUGHES, of Va.
GENERAL THOMAS. By HENRY COPPÉE, LL. D.
GENERAL SCOTT. By General MARCUS J. WRIGHT.
GENERAL WASHINGTON. By Gen. BRADLEY T. JOHNSON.
GENERAL LEE. By General FITZHUGH LEE.
GENERAL HANCOCK. By General FRANCIS A. WALKER.
GENERAL SHERIDAN. By General HENRY E. DAVIES.

These are volumes of especial value and service to school libraries, either for reference or for supplementary reading in history classes. Libraries, whether public, private, or school, that have not already taken necessary action, should at once place upon their order-lists the GREAT COMMANDERS SERIES.

The following are in press or in preparation:
General Sherman. By General MANNING F. FORCE.
General Grant. By General JAMES GRANT WILSON.
Admiral Porter. By JAMES F. SOLEY, late Assistant Sec'y of Navy.
General McClellan. By General ALEXANDER S. WEBB.
General Meade. By RICHARD MEADE BACHE.

"This series of books promises much, both by their subjects and by the men who have undertaken to write them. They are just the reading for young men and women; delightful reading for men and women of any age."—*The Evangelist.*

New York: D. APPLETON & CO., 72 Fifth Avenue.

D. APPLETON & CO.'S PUBLICATIONS.

JOHN BACH MCMASTER.

HISTORY OF THE PEOPLE OF THE UNITED STATES,

from the Revolution to the Civil War. By JOHN BACH MCMASTER. To be completed in six volumes. Vols. I, II, III, and IV now ready. 8vo. Cloth, gilt top, $2.50 each.

"... Prof. McMaster has told us what no other historians have told. . . . The skill, the animation, the brightness, the force, and the charm with which he arrays the facts before us are such that we can hardly conceive of more interesting reading for an American citizen who cares to know the nature of those causes which have made not only him but his environment and the opportunities life has given him what they are."—*N. Y. Times.*

"Those who can read between the lines may discover in these pages constant evidences of care and skill and faithful labor, of which the old-time superficial essayists, compiling library notes on dates and striking events, had no conception; but to the general reader the fluent narrative gives no hint of the conscientious labors, far-reaching, world-wide, vast and yet microscopically minute, that give the strength and value which are felt rather than seen. This is due to the art of presentation. The author's position as a scientific workman we may accept on the abundant testimony of the experts who know the solid worth of his work; his skill as a literary artist we can all appreciate, the charm of his style being self-evident."—*Philadelphia Telegraph.*

"The third volume contains the brilliantly written and fascinating story of the progress and doings of the people of this country from the era of the Louisiana purchase to the opening scenes of the second war with Great Britain—say a period of ten years. In every page of the book the reader finds that fascinating flow of narrative, that clear and lucid style, and that penetrating power of thought and judgment which distinguished the previous volumes."—*Columbus State Journal.*

"Prof. McMaster has more than fulfilled the promises made in his first volumes, and his work is constantly growing better and more valuable as he brings it nearer to our own time. His style is clear, simple, and idiomatic, and there is just enough of the critical spirit in the narrative to guide the reader."—*Boston Herald.*

"Take it all in all, the History promises to be the ideal American history. Not so much given to dates and battles and great events as in the fact that it is like a great panorama of the people, revealing their inner life and action. It contains, with all its sober facts, the spice of personalities and incidents, which relieves every page from dullness."—*Chicago Inter-Ocean.*

"History written in this picturesque style will tempt the most heedless to read. Prof. McMaster is more than a stylist; he is a student, and his History abounds in evidences of research in quarters not before discovered by the historian."—*Chicago Tribune.*

"A History *sui generis* which has made and will keep its own place in our literature."—*New York Evening Post.*

"His style is vigorous and his treatment candid and impartial."—*New York Tribune.*

New York: D. APPLETON & CO., 72 Fifth Avenue.

D. APPLETON & CO.'S PUBLICATIONS.

ACTUAL AFRICA; or, The Coming Continent. A Tour of Exploration. By FRANK VINCENT, author of "The Land of the White Elephant," etc. With Map and 102 Illustrations. 8vo. Cloth, $5.00.

This thorough and comprehensive work furnishes a survey of the entire continent, which this experienced traveler has circumnavigated in addition to his inland explorations. The latter have included journeys in northern Africa, Madagascar, southern Africa, and an expedition into the Congo country which has covered fresh ground. His book has the distinction of presenting a comprehensive summary, instead of offering an account of one special district. It is more elaborately illustrated than any book upon the subject, and contains a large map carefully corrected to date.

"Mr. Frank Vincent's books of travel merit to be ranked among the very best, not only for their thoroughness, but for the animation of their narrative, and the skill with which he fastens upon his reader's mind the impression made upon him by his voyagings."—*Boston Saturday Evening Gazette.*

"A new volume from Mr. Frank Vincent is always welcome, for the reading public have learned to regard him as one of the most intelligent and observing of travelers."—*New York Tribune.*

AROUND AND ABOUT SOUTH AMERICA: Twenty Months of Quest and Query. By FRANK VINCENT. With Maps, Plans, and 54 full-page Illustrations. 8vo, xxiv + 473 pages. Ornamental cloth, $5.00.

"South America, with its civilization, its resources, and its charms, is being constantly introduced to us, and as constantly surprises us. . . . The Parisian who thinks us all barbarians is probably not denser in his prejudices than most of us are about our Southern continent. We are content not to know, there seeming to be no reason why we should. Fashion has not yet directed her steps there, and there has been nothing to stir us out of our lethargy. . . . Mr. Vincent observes very carefully, is always good-humored, and gives us the best of what he sees. . . . The reader of his book will gain a clear idea of a marvelous country. Maps and illustrations add greatly to the value of this work."—*New York Commercial Advertiser.*

"The author's style is unusually simple and straightforward, the printing is remarkably accurate, and the forty-odd illustrations are refreshingly original for the most part."—*The Nation.*

"Mr. Vincent has succeeded in giving a most interesting and valuable narrative. His account is made doubly valuable by the exceptionally good illustrations, most of them photographic reproductions. The printing of both text and plates is beyond criticism."—*Philadelphia Public Ledger.*

IN AND OUT OF CENTRAL AMERICA; and other Sketches and Studies of Travel. By FRANK VINCENT. With Maps and Illustrations. 12mo. Cloth, $2.00.

"Few living travelers have had a literary success equal to Mr. Vincent's."—*Harper's Weekly.*

"Mr. Vincent has now seen all the most interesting parts of the world, having traveled, during a total period of eleven years, two hundred and sixty-five thousand miles. His personal knowledge of man and Nature is probably as varied and complete as that of any person living."—*New York Home Journal.*

New York: D. APPLETON & CO., 72 Fifth Avenue.

www.ingramcontent.com/pod-product-compliance
Lightning Source LLC
Chambersburg PA
CBHW021152230426
43667CB00006B/365